Immunosuppression and Immunomodulation

Edited by Rajeev K. Tyagi, Prakriti Sharma and Praveen Sharma

Published in London, United Kingdom

Immunosuppression and Immunomodulation
http://dx.doi.org/10.5772/intechopen.102203
Edited by Rajeev K. Tyagi, Prakriti Sharma and Praveen Sharma

Contributors
Miteshkumar Rajaram Maurya, Renuka Munshi, Sachin Bhausaheb Zambare, Sanket Thakur, Audrey Y-H. Teh, Julian K-C. Ma, Nicole Falci Finardi, Ibrahim Mustafa, Pratima Tripathi, Jacob George, Girish Vasudeo Kumthekar, Rajasekara Chakravarthi, Gatien A.G. Lokossou, Maximin Senou, Poppy Diah Palupi, Mohammed Safwan Ali Khan, Nur Dina Amalina, Ayda Cherian, Velmurugan Vadivel, Shumaila Nida Muhammad Hanif, Anna Hargrave, Abu Salim Mustafa, Javed Tunio, Asma Hanif

First published in London, United Kingdom, 2023 by IntechOpen
IntechOpen is the global imprint of INTECHOPEN LIMITED, registered in England and Wales, registration number: 11086078, 5 Princes Gate Court, London, SW7 2QJ, United Kingdom

British Library Cataloguing-in-Publication Data
A catalogue record for this book is available from the British Library

Additional hard and PDF copies can be obtained from orders@intechopen.com

Immunosuppression and Immunomodulation
Edited by Rajeev K. Tyagi, Prakriti Sharma and Praveen Sharma
p. cm.
Print ISBN 978-1-83768-071-9
Online ISBN 978-1-83768-072-6
eBook (PDF) ISBN 978-1-83768-073-3

Meet the editors

Dr. Rajeev K. Tyagi obtained a Ph.D. from the Biomedical Parasitology Unit, Institute Pasteur, France, in 2011 with a study on malaria immunology/parasitology. He developed a long-lasting, stable, and straightforward laboratory animal model to study the biology and immunology of infectious diseases and more. Dr. Tyagi continued to work on parasite immunology and mouse–human chimeras development at the University of South Florida, USA, and explored the asexual blood and liver stage infection of *P. falciparum*. Dr. Tyagi discovered a novel dendritic-like cell population called "pathogen differentiated dendritic cells (PDDCs)" when incubated with *P. gingivalis* and tracking of monocyte-derived dendritic cells (MoDcs) in a reconstituted immunodeficient NOD.$Prkdc^{scid}Il2rg^{-/-}$ (NSG) mice at Augusta University, USA. Dr. Tyagi explored IL-23R in the modulation of the functioning of regulatory T cells and its role in the pathogenesis of colitis in an experimental humanized mouse at Vanderbilt University Medical Centre (VUMC), USA. Currently, Dr. Tyagi leads a research group at CSIR-Institute of Microbial Technology, India.

Prakriti Sharma obtained an MSc in Biotechnology from HNB Gharwal University, India, in 2013. He studied the characterization and diversity of plant growth-promoting rhizobacteria from maize. Dr. Sharma obtained a Ph.D. in Biotechnology from Guru Angad Dev Veterinary and Animal Sciences University, India, in 2022 with in vivo studies on the role of nuclear factor of activated T-cell (NFAT) signaling pathway in lung damage following exposure to deltamethrin insecticide.

Dr. Praveen Sharma obtained a Ph.D. from the Laboratory of Molecular Medicine, Central University of Punjab, India, in 2021 with a study on the regulation of mitochondrial metabolism mediated by small regulatory molecules. Dr. Sharma works on developing a further understanding of the molecular aspects of breast cancer pathology. Additionally, Dr. Sharma has worked in collaboration with pharmacological laboratories to investigate the anti-cancer effects of different synthesized molecules and various natural extracts.

Contents

Preface

Immunosuppression and immunomodulation are different but related terms. Immunosuppression is the suppression of the immune system and immunomodulation is the alteration of immune response due to, for example, infections or chemical exposures. These processes can also be activated by drugs and chemicals being used to treat immune system-related health problems. This book provides a comprehensive overview of immunosuppression and immunomodulation. It discusses the pros and cons of these processes as well as the various natural and synthetic agents that lead to immune problems. Furthermore, this book discusses the various methods and drugs used for immunosuppression and immunomodulation in cancer therapy, rheumatoid arthritis treatment, organ transplantation, and more. This volume is a useful resource for students and researchers studying the immunology of infectious and inflammatory diseases.

Rajeev K. Tyagi, Ph.D. and Prakriti Sharma, Ph.D.
Division of Cell Biology and Immunology,
Biomedical Parasitology and Translational-immunology Lab,
CSIR-Institute of Microbial Technology (IMTECH),
Chandigarh, India

Dr. Praveen Sharma, Ph.D.
Laboratory of Molecular Medicine,
Central University of Punjab,
Bathinda, India

Section 1

Introduction

Chapter 1

Introductory Chapter: *P. falciparum* Modulates Dendritic Cell Functions to Circumvent the Host Immune Response

Nikunj Tandel, Mansi Thakkar, Prakriti Sharma and Rajeev K. Tyagi

1. Introduction

Malaria, a human parasite infectious disease, has been a cause of mortality and morbidity across the world. Significant advancements toward the vaccine development have been made, yet half of the world's population survives under the threat of malaria infection, particularly young children residing in South East Asia (SEA) and Sub-Saharan Africa. As per the latest World Malaria Report 2022, 247 million cases of malaria were registered in 2021, slightly higher as compared to 245 million in 2020. Additionally, malaria death increased by 10% and reached an estimated number of 6,25,000 [1]. Malaria occurs due to the bite of female *Anopheles* mosquito which carries the infectious sporozoites of *Plasmodium* species. According to geographical location and environmental conditions, from the range of *Plasmodium* species, there are mainly four species that cause malaria infection in humans; *P. falciparum, vivax, malariae, and ovale* [2]. *P. falciparum* is the most fatal and leading cause of death in humans. Also, it leads to the development of cerebral malaria. *P. vivax* remains in the dormant stage for a prolonged period and develops an infection in the later stages.

The tens of millions of non-immunes from areas where malaria is not transmitted visit malaria-endemic areas, and face risks of malaria infection. The two major weapons against malaria are vector control and chemoprophylaxis/chemotherapy. Unfortunately, attempts to eradicate the disease based on these methods have had only limited success due to widespread development of drug resistance by the parasite and insecticide resistance by the mosquito vector [3]. Therefore, there is an urgent need to develop newer drugs and therapeutic approaches. There is no single effective malaria vaccine available due to the complex life cycle of malaria parasite; a number of approaches to malaria vaccine development based on attenuated sporozoite, synthetic and recombinant immunogenic peptide is available. However, these approaches suffer from the drawbacks of safety and short-lived species & stage-specific immunity [3].

There are antimalarial drug(s) available to treat human malaria infection, but continuous drug pressure to clear *P. falciparum* led to the development of drug pressure endurance to tolerate the therapeutic effects of the drugs. There have been many mechanisms of action that parasites may have employed to escape the drug pressure. The emergence of resistance against frontline antimalarials and their combinations by *P. falciparum* is worrisome as it threatens to make malaria practically untreatable in SEA. This will have severe implications as it would hinder the global attempts to eliminate this deadliest human disease. A recent series of clinical trials, *in vitro*, genomics, and transcriptomic studies in SEA have defined *in vivo* and *in vitro* phenotypes of artemisinin resistance; identified its causal genetic determinant; explored its molecular mechanism; and assessed its clinical impact [4]. The artemisinin-based combination therapy (ACT) is the only remaining remedy to clear the parasite infection. However, tolerance shown by the parasite toward the combination of drugs and issue of co-resistance led researchers to develop an understanding of how do parasites escape the therapeutic effect of drugs.

Malaria life cycle begins with the bite of *Anopheles* mosquito which transmits the infectious motile sporozoites. Once it enters the human host, by escaping the host immune system it reaches the liver and develops into the liver stage. Following propagation in the liver, it comes out into the bloodstream where its prime targets are the circulating red blood cells (RBCs). Subsequently, these infected RBCs (iRBCs) further infect other healthy RBCs and result in the development of blood-stage infection which showcases the symptoms of fever, shivering, and others. During this continuous cycle, certain iRBCs convert into the male and female gametocytes. These sexual forms of parasites are taken up by the mosquito during the biting and further develop into the mosquito gut followed by becoming sporozoites and reside in the salivary glands of the mosquito and are further injected into the healthy human. With the progression of time, studies have revealed the different stages of malaria infection (in human and mosquito host) which helps in understanding the host-pathogen interaction. The malaria life cycle of *P. falciparum* has been shown in **Figure 1**.

2. Modulation of host immune system

The malaria life cycle in the human host initiates in the liver followed by symptomatic blood stage infection. Different experimental studies of humans and mice have confirmed the role of immune system to fight against the infection [2]. Further, studies have shown the importance of T cells, mainly IFN-γ producing $CD8^+$ T cells which have a prominent role in providing sterile protection during the infectious challenge. Moreover, other cells such as IFN-γ producing $CD4^+$ T cells and follicular helper T cells (T_{fh}) also play an important role in killing iRBCs and generation of antibody-producing B cells, respectively [2]. Additionally, different immune cells of innate immunity also support augmenting the immune response to the malaria infection [2]. It has been well-established that among the different immune cells, distinct mononuclear phagocytic cells known as dendritic cells (DCs) are consider as professional antigen-presenting cells (APCs). DCs have been well-known APCs to identify antigens, capturing, processing, and presentation to the T cells as well as activating B cells directly [6]. Furthermore, it also stimulates the innate immune system (activation of NK cells). The role of DCs is well-defined. These cells work in coordination with other immune cells and bridge the gap between adaptive and non-adaptive immunity (**Figure 2**).

They are classified into different subsets and majority of them are divided according to the expression of certain defined phenotypic markers and location [6]. They reside in lymphoid and non-lymphoid organs and mainly classified as conventional/myeloid dendritic cells (cDCs/mDCs) ($CD3^-CD14^-CD19^-CD20^-CD56^-HLA\text{-}DR^+DC11c^+$)

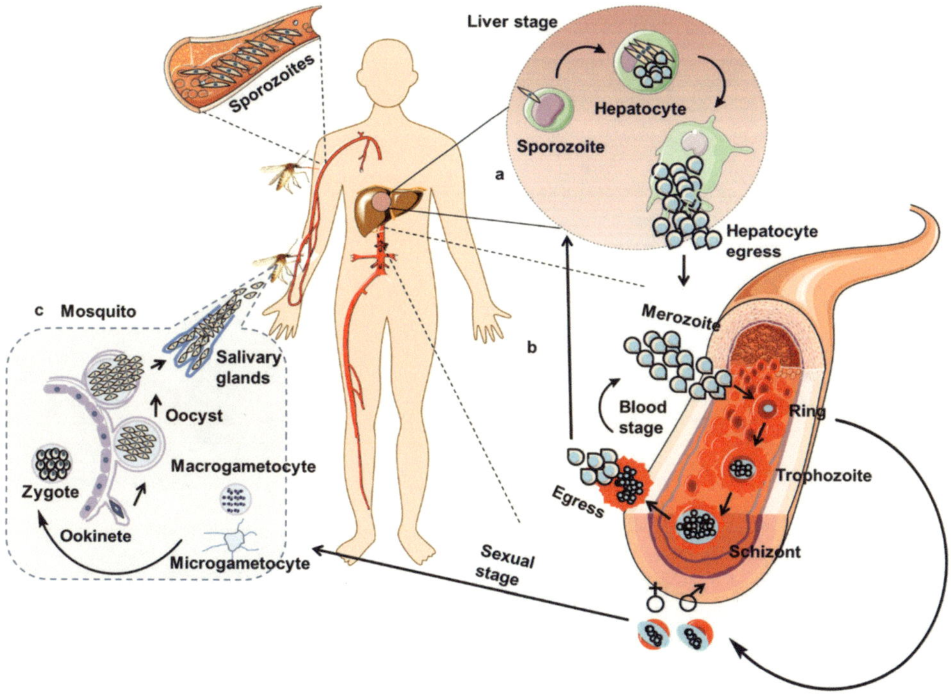

Figure 1.
Complex life cycle of P. falciparum. The life cycle has 3 stages: The pre-erythrocytic and erythrocytic stages in humans (host) and the sexual process in the mosquito vector a) pre-erythrocytic stage b) Erythrocytic stage, and c) mosquito/sexual stage. (adapted with permission from [5]).

and plasmacytoid dendritic cells (pDCs) ($CD3^-CD14^-CD19^-CD20^-CD56^-HLA\text{-}DR^+DC11c^+CD303(BDCA2)^+CD304(BDCA4)^+$). These mDCs in blood and lymphoid tissues can be further divided into two more subsets which express CD1c (BDCA1) or CD141 (BDCA3). pDCs are the major reservoir for antiviral immune response (IFN-α) which consist of 0.35% of PBMC whereas cDCs have a captive role in priming of T cells and account for 0.65% of PBMCs [7].

It has been established that protection against the malaria infection (liver/blood stage infection) is initiated when the DCs or macrophages capture the malaria antigen followed by processing and presenting them to T cells through MHC-I or II pathway. During the antigen presentation, several signaling mechanisms resulted in the secretion of pro-inflammatory cytokines such as IFN-γ, IL-12, and TNF-α. It further activates/stimulates the other immune cells and results in the direct/indirect killing of infected cells. Experimental studies have confirmed that DCs play a dual role by producing cytokines against the respective pathogen and creating tolerogenic conditions [6, 7].

2.1 Functional DCs during malaria infection

Role of DCs during malaria infection has been recently reviewed showing contradictory outcomes. However, it could be reasoned due to the use of different species and a subset of DCs. Therefore, our main focus is on understanding the mechanism of *P. falciparum* that modulates the DCs function and results in the exacerbated infection. Further, it is proven that interaction between DCs and *Plasmodium* parasites

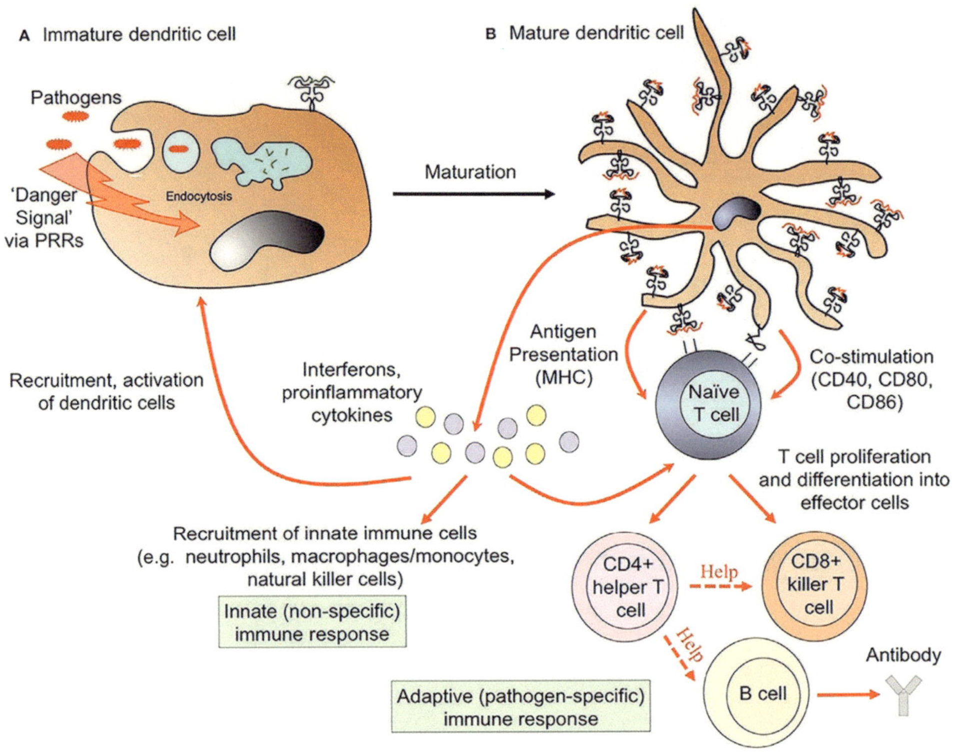

Figure 2.
Dendritic cells link innate and adaptive arms of the immune system (a) uptake of pathogens and recognition of pathogen-associated "danger signals" by pattern recognition receptors (PRRs) triggers dramatic morphological and functional changes in DCs, termed maturation. These changes involve the formation of dendrites, downregulation of antigen uptake, and redistribution of the major histocompatibility complex (MHC) molecules from intracellular endocytic compartments to the cell surface (b) mature DCs migrate to draining lymph nodes and present information about the invading pathogen in the form of processed peptides loaded onto MHC molecules to naïve T cells. Upregulation of MHC and co-stimulation molecules enables activated DCs to initiate adaptive T and B cell immune responses, the nature of which is determined by the cytokine milieu. This initiates the cascade to an adaptive immune response, leading to the clearance of infected cells, and the extracellular pathogens. Activated mature DCs also secrete interferons and pro-inflammatory cytokines that recruit circulating innate immune cells to provide rapid defense against infection (adapted with permission from [7]).

occurs at each stage of infection in human host led by the spleen and blood [7]. Further, tissue-residential DCs also have the capacity to phagocyte the parasitic components and generate the adaptive immune response. However, DCs in all these tissues have different levels of maturation and varying activation and generation of adaptive and innate immunity. DCs play a tolerogenic role to halt the immunopathology of infection in liver, whereas in peripheral blood it provokes an intermediate immune response than the potent responses seen in spleen [8].

DCs have a different role to play based on location specificity. The liver resident DCs are less mature and express lower costimulatory molecules compared to the blood DCs which accounts for their poor antigen presentation [9]. Additionally, their allogenic T cell response is also lower compared to their blood counterpart which results in less T cell-based response against the subsequent stimulation [10]. Whereas liver DCs are prominent IL-10-producing cells which favor the survival of sporozoites in the liver and hence fail to generate sterile immunity against natural infection [9]. Thus, after successfully invading the immune system by marginally around 30% of sporozoites, the tolerogenic nature of liver DCs is the first step for the development of malaria infection which could further progress and develop immunopathology. In this context, developed humanized mice may be a valuable tool to explore and study the role of DCs in liver-stage malaria infection [11].

Once the malaria infection reaches to blood stage, it allows the host immune system to activate and respond accordingly as a range of innate and toll-like receptors (TLRs) get activated. *P. falciparum* infection attacks the mature RBCs which do not express surface MHC and hence support the invasion from the host-immune response. Once the parasite matures in iRBCs they rupture and release thousands of merozoites in the circulation. If merozoites fail to infect healthy RBCs, they could be phagocytosed or reached to the spleen for clearance [12]. Later on, it was shown that *Pf*EMP1 molecule, expresses on iRBCs, plays a dual role during the blood-stage infection. It is the prime source for antibodies in the initial stage of infection whereas a study shows that it modulates the expression of CD36 by binding on APCs including DCs [13]. Further, it has shown that the modulated DCs have the capacity to express TNF-α, yet it fails to activate T cells and produce IL-10.

Role of DCs in human malaria infection has been studied mainly in two ways. One in which peripheral DCs of infected or pre-and-post infection DCs and in another way *in-vitro* isolated and stimulated DCs (via IL-4 and GM-CSF) and their interaction with different *Plasmodium* stimuli (pRBCs, synthetic hemozoin, *P. falciparum* merozoites) [7].

2.1.1 DCs and P. falciparum interaction

P. falciparum being the reason for greater mortality and morbidity, mainly targets children and pregnant women leading to death if goes undiagnosed and not treated. The studies conducted on children (in Kenya) depicted that irrespective of disease severity, it mainly reduces the expression of HLA-DR on cDCs not on pDCs alongside DCs numbers [14, 15]. Later on, it was found that increasing infection directly correlated with an elevated number of BDCA-3^{+}cDC1s. Interestingly, this effect persists for around two weeks after the discharge of patients suggesting that despite the clearance of parasites the immunosuppressive effect has not weakened [15]. Similar results were found when the study was conducted on children of 2–10 years in Mali between infected and non-infected once. The reduced expression of HLA-DR, an elevated number of BDCA-2^{+}pDC1 and BDCA-3^{+}cDC1, and less expression of CD86 were noticed after the malaria infection [16]. Furthermore, study carried out by Guermonprez and colleagues has also found similar results about the increased number of BDCA-3^{+}cDCs [17]. Other studies confirmed that it correlates with enhanced serum DCs growth factor, Flt3-L (Fms-like tyrosine kinase receptor 3 ligand), which provokes cDC1 and pDCs *in vivo* [18, 19]. This receptor is a product of uric acid metabolism driven by *Plasmodium* species and generated by the mast cells.

Studies conducted on DCs role in pregnant women have shown contradictory results. Out of four studies, two studies have shown the overall decrease in DCs population in *P. falciparum*-infected pregnant women in comparison to the uninfected pregnant women [20, 21]. Whereas a study conducted by Mamadou and colleagues only observed a decrease in the number of pDCs in infected once [22]. However, study of Fievet *et al.*, on Beninese pregnant infected/uninfected women did not see any alteration in the number of DCs [23]. The discrepancies in the results could be due to several reasons such as gating strategies and source of DCs, stages of pregnancy, and inclusion of controls. Studies conducted on Papua, Thailand, and Brazilian people have shown the role of DCs in low-transmission settings. The reduced circulating pDCs were observed in both types of infected people (mild and severe) [24].

Human studies confirmed that functional impairment of DCs in malaria is common. The endemic and higher transmission showed the parasite load and higher chances of re-infection. Furthermore, studies of co-infection with two *Plasmodium* species showed similar results and overall reduction in DCs in peripheral blood [25, 26]. The correlation between malaria infection and impairment functional activity of DCs is yet to be established. Interestingly, expression of HLA-DR on DCs was positively correlated with parasitemia in children having asymptomatic *P.vivax* infection, whereas negatively associated with parasitemia in adults having asymptomatic *P. falciparum* infection. Based on these data, distinct mechanism played by individual parasites and age-factor does play a dominant role in endemic area for considering the risk factor.

2.1.2 P. falciparum modulates TLRs present on DCs

The study conducted by Loharungsikul and colleagues has detailed the role of *P. falciparum* modulation of TLRs on DCs [24]. The comparison of infected (mild/severe) and non-infected people has shown the reduction of TLR9 expression on pDCs, whereas increased TLR2 expression on cDCs was seen. Additionally, no changes were observed in TLR4 expression [24]. However, there was no evidence found for the correlation between disease severity and alteration in TLR expression. Experimental studies have proven the important role of TLR2, 4, and 9 in sensing the *Plasmodium* species. It was confirmed later on that TLR 2 and 4 help recognize the GPI anchor for merozoite surface proteins [27], and TLR9 is known to identify the DNA of *Plasmodium* [28]. Detailed investigations are indeed needed to confirm the role played by TLRs in modulating the host response by the alteration of DCs function.

2.1.3 In vitro modulation of DCs

Earlier studies have shown the role of TLRs during malaria infection. To study more in detail, isolated peripheral blood mononuclear cells (PBMCs) of pregnant women (naturally exposed to *P. falciparum*) were stimulated with TLR3 (poly (I:C)), TLR4 (LPS) and TLR9 (CpG-A ODN) ligand [23]. Additionally, they were also stimulated with synthetic hemozoin products of haemin chloride. There was no difference in HLA-DR expression in infected or uninfected controls irrespective of any stimuli were seen. Whereas, enhanced production of TNF-α, IFN-γ, and IL-10 and TNF-α was measured against the PBMCs stimulated (of infected women) with hemozoin, poly(I:C), and CpG-A, respectively [23]. In another recent study, DCs were purified from the blood of adults residing in Mali and stimulated with parasitized red blood cells (pRBC) during the end season of transmission [29]. The stimulation of pRBCs with DCs (3:1 ratio) has upregulated the expression of CD86 and HLA-DR whereas expressed CXCL10, CCL2, and CXCL9. However, they failed to produce IL-10, TNF-α, IL-6, or IL-1β [29]. This data indicates that DCs can restore functional characteristics during the reduction in transmission.

Only fewer studies have been carried out using the *bonafide* population of DCs and the majority of the work done using BDCA-1$^+$ cDC2 and pDCs. Upregulation of co-stimulatory molecules alongside secretion of IFN-α during the incubation of peripheral DCs with pRBCs and merozoites suggest that *P. falciparum* has the capacity to induce naïve DCs [29–31]. The detailed analysis showed contradictory results with moDCs studies and controlled human malaria infection (CHMI) studies that pointed out the cross-talk between different populations of DCs in the generation of immune response against *P. falciparum* infection [7].

2.1.4 Modulation of DCs from controlled human malaria infection (CHMI) studies

The controlled human malaria infection model (CHMI) is one of the successful models developed for understanding host-pathogen interaction. This has provided us with significant insights into antimalarial immunity. Woodberry and colleagues carried out a study to understand the role of DCs in malaria by inoculating the ultra-low or low numbers of *P. falciparum* pRBCs [32]. Drug treatment was given on day 6 post-inoculation or parasitemia reached 1000/ml. They found the curtailed down population of DCs due to the apoptosis, mainly DCs expressing HLA-DR^{+}. It was also found to be correlated with symptomatic malaria. The number of cDCs was found to be recovered in comparison to the pDCs which remains at the base level of around 47% 60 hr. post-treatment [32]. Despite the overall recovery of DCs, phagocytic activity was found to be impaired after 36 hr. of treatment. Overall, this study depicts that a specific number of sporozoites are required for the functional impairment of DCs. However, treatment in the ultra-low dose group before any symptoms has raised the question about the said correlation.

In this direction, another two studies were done to study more about BDCA-1^{+} cDC2 activation [33] and the function of pDC [34]. The results of both studies were found to be similar. Moreover, elevated apoptosis in DCs with a reduction in number and its decreased phagocytic activity was found only in the higher-dose group. Additionally, Loughland *et al.*, have also analyzed the DCs population followed by TLR stimulation (TLR4, TLR7, and TLR1/2) [33]. They have seen that BDCA-1^{+} cDC2 population failed to express CD86 and HLA-DR after TLR stimulation whereas upregulation of IFN-γ, HLA-DR, and CD123 was observed on pDCs upon stimulation with TLR7 and TLR9. Consistent with earlier findings, studies were carried out on the stimulation of DCs with CpG-A isolated from pregnant women. In summary, it is concluded that even a small number of infected sporozoites may lead to impairment in the function of cDCs whereas not affecting the pDCs population.

2.1.5 Interaction between DCs and parasite-generated metabolic products

Parasite progresses inside the human host and mainly relies on the nutrient available in the vicinity. Hemoglobin, the major target of the parasite, is the key product as it metabolite and results into the formation of heme which is further neutralized by parasites and converted into hemozoin [35]. It plays a dual role in the activation and suppression of DCs. Later on, it has been confirmed that hemozoin serves as a carrier for *Plasmodium* DNA and presents to TLR9 [36]. Similarly, uric acid, a toxic product of *P. falciparum*, also accounts for the up-and-down regulation of co-stimulatory molecules (CD86, CD80, and CD11c) and HLA-DR, respectively of human peripheral DCs [37]. Later on, it has revealed that uric acid was responsible for the inflammation during *Plasmodium* infection by activating the inflammasome. However, its role in antimalarial DCs response needs to be investigated [38]. Collectively, these studies suggest that *P. falciparum* DNA is responsible for the activation of TLR ligands, especially TLR9 on pDCs as it is the only TLR ligand expressed by human pDCs.

3. Conclusions

The role of DCs in malaria infection failed to understand the immunopathology due to several factors in-and-out of experiments. Similarly, the studies which had a focus on direct interaction between *Plasmodium* species and DCs have used human monocyte-derived DCs (moDCs) due to their easy *ex-vivo* generation. However, recent studies have confirmed that moDCs are distinct from blood and cDCs populations and may not represent the true DCs population. Additionally, it has been also hypothesized that inflammatory moDCs are similar to the macrophages and not *bonafide* population of DCs. Further, the relationship between pRBCs and moDCs for their inhibition or activation and how particular stimuli play a cascade role in it is still elusive. Therefore, understanding how malaria infection modulates DCs functions followed by their suppression is not fully studied. Also, whether it can happen directly through interaction with DCs or the involvement of other mediators such as cytokines or metabolites play a crucial role warrants further detailed investigations.

Acknowledgements

Rajeev K. Tyagi would like to express his gratitude to DBT, New Delhi, Ramalingaswami Re-entry Fellowship Project (No. BT/RLF/Re-entry/27/2018) and Indian Council of Medical Research (ICMR), New Delhi extramural grant (35/1/2020-Nano/BMS) for generously supporting this study. Rajeev K. Tyagi would like to express his thanks to the central MIL facility of CSIR-IMTECH, Chandigarh. Nikunj Tandel would like to thank the Nirma University and the Indian Council of Medical Research (ICMR) for providing the fellowship to carry out his research (ICMR award letter No.: 2020-7623/CMB-BMS).

Conflict of interest

The authors declare no conflict of interest.

Abbreviations

ACT	artemisinin-based combination therapy
APC	antigen-presenting cells
cDCs/mDCs	conventional/myeloid dendritic cells
CHMI	controlled human malaria infection model
DCs	dendritic cells
Flt3-L	Fms-like tyrosine kinase receptor 3 ligand
iRBCs	infected red blood cells
MHC	major histocompatibility complex
pRBC	parasitized red blood
PPR	pattern recognition receptors
RBCs	red blood cells
SEA	South-East Asia
TLRs	toll-like receptors

Author details

Nikunj Tandel[1], Mansi Thakkar[1], Prakriti Sharma[2] and Rajeev K. Tyagi[2]*

1 Institute of Science, Nirma University, Ahmedabad, Gujarat, India

2 Division of Cell Biology and Immunology, Biomedical Parasitology and Nano-Immunology Lab, CSIR-Institute of Microbial Technology (IMTECH), Chandigarh, India

*Address all correspondence to: rajeevtyagi@imtech.res.in; rajeev.gru@gmail.com

References

[1] WHO. World Malaria Report 2022. Geneva, Switzerland: World Health Organization; 2022

[2] Nikunj T, Sarat KD. T cell-based vaccines: Hope for malaria elimination. In: Alfonso JR-M, editor. Current Topics and Emerging Issues in Malaria Elimination. IntechOpen: Rijeka; 2021. p. Ch. 15

[3] Tandel N, Tyagi RK. Chapter 5 - Malaria. In: Misra G, Srivastava V, editors. Molecular Advancements in Tropical Diseases Drug Discovery. London, UK: Academic press is an imprint of Elsevier; 2020. pp. 95-116

[4] Fairhurst RM, Dondorp AM. Artemisinin-Resistant Plasmodium falciparum Malaria. Microbiology spectrum. 2016;**4**:3

[5] Bonam SR, Rénia L, Tadepalli G, Bayry J, Kumar HMS. Plasmodium falciparum Malaria Vaccines and Vaccine Adjuvants. Vaccine. 2021;**9**(10):1072

[6] Amorim KN, Chagas DC, Sulczewski FB, Boscardin SB. Dendritic cells and their multiple roles during malaria infection. Journal of Immunology Research. 2016;**2016**:2926436

[7] Yap XZ, Lundie RJ, Beeson JG, O'Keeffe M. Dendritic cell responses and function in malaria. Frontiers in Immunology. 2019;**10**:357

[8] Mittag D, Proietto AI, Loudovaris T, Mannering SI, Vremec D, Shortman K, et al. Human dendritic cell subsets from spleen and blood are similar in phenotype and function but modified by donor health status. Journal of Immunology (Baltimore, Md. : 1950). 2011;**186**(11):6207-6217

[9] Bamboat ZM, Stableford JA, Plitas G, Burt BM, Nguyen HM, Welles AP, et al. Human liver dendritic cells promote T cell hyporesponsiveness. Journal of Immunology (Baltimore, Md. : 1950). 2009;**182**(4):1901-1911

[10] Bosma BM, Metselaar HJ, Mancham S, Boor PP, Kusters JG, Kazemier G, et al. Characterization of human liver dendritic cells in liver grafts and perfusates. Liver transplantation : official publication of the American Association for the Study of Liver Diseases and the International Liver Transplantation Society. 2006;**12**(3):384-393

[11] Tyagi RK, Tandel N, Deshpande R, Engelman RW, Patel SD, Tyagi P. Humanized mice are instrumental to the study of plasmodium falciparum infection. Frontiers in Immunology. 2018;**9**:2550

[12] Trubowitz S, Masek B. Plasmodium falciparum: Phagocytosis by polymorphonuclear leukocytes. Science (New York, N.Y.). 1968;**162**(3850):273-274

[13] Urban BC, Willcox N, Roberts DJ. A role for CD36 in the regulation of dendritic cell function. Proceedings of the National Academy of Sciences of the United States of America. 2001;**98**(15):8750-8755

[14] Urban BC, Mwangi T, Ross A, Kinyanjui S, Mosobo M, Kai O, et al. Peripheral blood dendritic cells in children with acute plasmodium falciparum malaria. Blood. 2001;**98**(9):2859-2861

[15] Urban BC, Cordery D, Shafi MJ, Bull PC, Newbold CI, Williams TN, et al. The frequency of BDCA3-positive dendritic cells is increased in the peripheral circulation of Kenyan children

with severe malaria. Infection and Immunity. 2006;**74**(12):6700-6706

[16] Arama C, Giusti P, Boström S, Dara V, Traore B, Dolo A, et al. Interethnic differences in antigen-presenting cell activation and TLR responses in Malian children during plasmodium falciparum malaria. PLoS One. 2011;**6**(3):e18319

[17] Guermonprez P, Helft J, Claser C, Deroubaix S, Karanje H, Gazumyan A, et al. Inflammatory Flt3l is essential to mobilize dendritic cells and for T cell responses during plasmodium infection. Nature Medicine. 2013;**19**(6):730-738

[18] Maraskovsky E, Brasel K, Teepe M, Roux ER, Lyman SD, Shortman K, et al. Dramatic increase in the numbers of functionally mature dendritic cells in Flt3 ligand-treated mice: Multiple dendritic cell subpopulations identified. The Journal of Experimental Medicine. 1996;**184**(5):1953-1962

[19] Waskow C, Liu K, Darrasse-Jèze G, Guermonprez P, Ginhoux F, Merad M, et al. The receptor tyrosine kinase Flt3 is required for dendritic cell development in peripheral lymphoid tissues. Nature Immunology. 2008;**9**(6):676-683

[20] Ibitokou S, Oesterholt M, Brutus L, Borgella S, Agbowaï C, Ezinmègnon S, et al. Peripheral blood cell signatures of plasmodium falciparum infection during pregnancy. PLoS One. 2012;7(12):e49621

[21] Breitling LP, Fendel R, Mordmueller B, Adegnika AA, Kremsner PG, Luty AJ. Cord blood dendritic cell subsets in African newborns exposed to plasmodium falciparum in utero. Infection and Immunity. 2006;**74**(10):5725-5729

[22] Diallo M, Aldebert D, Moreau JC, Ndiaye M, Jambou R. Decrease of lymphoid dendritic cells in blood from malaria-infected pregnant women. International Journal for Parasitology. 2008;**38**(13):1557-1565

[23] Fievet N, Varani S, Ibitokou S, Briand V, Louis S, Perrin RX, et al. Plasmodium falciparum exposure in utero, maternal age and parity influence the innate activation of foetal antigen presenting cells. Malaria Journal. 2009;**8**:251

[24] Loharungsikul S, Troye-Blomberg M, AmoudruzP, PichyangkulS, YongvanitchitK, Looareesuwan S, et al. Expression of toll-like receptors on antigen-presenting cells in patients with falciparum malaria. Acta Tropica. 2008;**105**(1):10-15

[25] Kho S, Marfurt J, Noviyanti R, Kusuma A, Piera KA, Burdam FH, et al. Preserved dendritic cell HLA-DR expression and reduced regulatory T cell activation in asymptomatic plasmodium falciparum and P. vivax infection. Infection and Immunity. 2015;**83**(8):3224-3232

[26] Kho S, Marfurt J, Handayuni I, Pava Z, Noviyanti R, Kusuma A, et al. Characterization of blood dendritic and regulatory T cells in asymptomatic adults with sub-microscopic plasmodium falciparum or plasmodium vivax infection. Malaria Journal. 2016;**15**:328

[27] Nebl T, De Veer MJ, Schofield L. Stimulation of innate immune responses by malarial glycosylphosphatidylinositol via pattern recognition receptors. Parasitology. 2005;**130**(Suppl):S45-S62

[28] Coban C, Ishii KJ, Kawai T, Hemmi H, Sato S, Uematsu S, et al. Toll-like receptor 9 mediates innate immune activation by the malaria pigment hemozoin. The Journal of Experimental Medicine. 2005;**201**(1):19-25

[29] Götz A, Tang MS, Ty MC, Arama C, Ongoiba A, Doumtabe D, et al. Atypical activation of dendritic cells by plasmodium falciparum. Proceedings of the National Academy of Sciences of the United States of America. 2017;**114**(49):E10568-e10577

[30] Pichyangkul S, Yongvanitchit K, Kum-arb U, Hemmi H, Akira S, Krieg AM, et al. Malaria blood stage parasites activate human plasmacytoid dendritic cells and murine dendritic cells through a toll-like receptor 9-dependent pathway. Journal of immunology (Baltimore, Md. : 1950). 2004;**172**(8):4926-4933

[31] Wu X, Gowda NM, Kumar S, Gowda DC. Protein-DNA complex is the exclusive malaria parasite component that activates dendritic cells and triggers innate immune responses. Journal of immunology (Baltimore, Md. : 1950). 2010;**184**(8):4338-4348

[32] Woodberry T, Minigo G, Piera KA, Amante FH, Pinzon-Charry A, Good MF, et al. Low-level plasmodium falciparum blood-stage infection causes dendritic cell apoptosis and dysfunction in healthy volunteers. The Journal of Infectious Diseases. 2012;**206**(3):333-340

[33] Loughland JR, Minigo G, Burel J, Tipping PE, Piera KA, Amante FH, et al. Profoundly reduced CD1c+ myeloid dendritic cell HLA-DR and CD86 expression and increased tumor necrosis factor production in experimental human blood-stage malaria infection. Infection and Immunity. 2016;**84**(5):1403-1412

[34] Loughland JR, Minigo G, Sarovich DS, Field M, Tipping PE, Montes de Oca M, et al. Plasmacytoid dendritic cells appear inactive during sub-microscopic plasmodium falciparum blood-stage infection, yet retain their ability to respond to TLR stimulation. Scientific Reports. 2017;**7**(1):2596

[35] Boura M, Frita R, Góis A, Carvalho T, Hänscheid T. The hemozoin conundrum: Is malaria pigment immune-activating, inhibiting, or simply a bystander? Trends in Parasitology. 2013;**29**(10):469-476

[36] Parroche P, Lauw FN, Goutagny N, Latz E, Monks BG, Visintin A, et al. Malaria hemozoin is immunologically inert but radically enhances innate responses by presenting malaria DNA to toll-like receptor 9. Proceedings of the National Academy of Sciences of the United States of America. 2007;**104**(6):1919-1924

[37] van de Hoef DL, Coppens I, Holowka T, Ben Mamoun C, Branch O, Rodriguez A. Plasmodium falciparum-derived uric acid precipitates induce maturation of dendritic cells. PLoS One. 2013;**8**(2):e55584

[38] Gallego-Delgado J, Ty M, Orengo JM, van de Hoef D, Rodriguez A. A surprising role for uric acid: The inflammatory malaria response. Current Rheumatology Reports. 2014;**16**(2):401

Section 2

Immunity and Immunouppression

Chapter 2

Immune Response to HIV-1 Infection and Vaccine Development

Anna Hargrave, Abu Salim Mustafa, Asma Hanif, Javed Tunio and Shumaila Nida Muhammad Hanif

Abstract

Human immunodeficiency virus (HIV)-1 infection represents an ongoing challenging public health epidemic. This is in part because of the socioeconomic burden on low-income countries, lack of access to highly active antiretroviral therapy and other medical treatment, and progression to acquired immunodeficiency syndrome (AIDS) over the course of years. To control or eradicate this virus, a prophylactic vaccine must be generated. Despite several decades of research, development, and clinical trials, there is not yet an effective immunization. This chapter focuses on unique aspects of the immune response to this infection, challenges of vaccine development, key clinical trials, and promising vaccine strategies.

Keywords: HIV-1 infection, prophylactic vaccine, immune response to HIV-1, vaccine development, promising vaccine strategies

1. Introduction

Currently, the UNAIDS estimates that there are 38.4 million people with human immunodeficiency virus (HIV)-1 infections worldwide as of 2021. Of these 38.4 million people, approximately a quarter of people (25.2% or 9.7 million) with HIV-1 infections are not accessing the standard of care, antiretroviral therapy (ART) [1].

ART is the only known treatment to slow the progression to acquire immunodeficiency syndrome (AIDS) and to prevent the spread of infection, but there are several shortcomings with this therapy, as discussed in further detail in the next section.

HIV-1 infection and its progression to AIDS is a global health crisis and disproportionally affects low-income countries, particularly those in Sub-Saharan Africa. Though there is 12% of the global population in Sub-Saharan Africa, this region has 71% of the world's population of people with HIV infections [2]. HIV infection is the highest in young men and women in child-bearing age [3], with women having up to eightfold higher risk for acquiring HIV than their male cohorts [2]. These women with ages between 15 and 24 are also more likely to acquire the infection five to seven years before men in the same age group [2]. Given the great number of women impacted by HIV-1 infections, this leads to

 IntechOpen

pediatric infections from perinatal transmission or transmissions from breast milk. Notably, pediatric infections are very different than adult infections and detailed in Section 5.

It is widely accepted that immunization is the most cost-effective, scalable, and lasting public health intervention method to end the HIV epidemic [4]. Though a prophylactic HIV-1 vaccine has been researched and developed for several decades, there is yet to be a licensed vaccine on the market. The main obstacles preventing the generation of a successful vaccine are as follows: (i) HIV-1 has a high rate of mutation and viral replication, with extraordinary worldwide genetic diversity [4], as discussed in the following paragraph and "immune response" section, (ii) immune response behind an HIV-1 infection is never completely eradicated, leading to an incomplete understanding of correlates of immune protection [4], (iii) HIV-1 creates a latent reservoir unlike other viral infections, as reviewed in Section 3, (iv) there is no appropriate animal model [4], and (v) there are funding issues associated with vaccine development [4, 5].

2. Limitations of ART

Despite its significant role in the prevention of HIV-1 infection transmission, there are several social, economic, and technical obstacles associated with ART therapy.

Besides coping with social stigma for ART [6], there is the social and economic challenge of the limited availability of ART therapy in low-income countries [7]. Given that the countries most affected by HIV are resource poor, only a small proportion of people of the global population with HIV have benefited ART [3]. Despite scaling up ART in several of these countries of sub-Saharan Africa, there are still countries where less than 25% of the adult population has access to ART [2]. Additionally, there is limited screening, so people may be unaware of their positive HIV status and never enroll in ART therapy [2].

The technical shortcomings of ART include the following: (i) limited impact on latent reservoir, (ii) strict, daily adherence to drug regimen, (iii) requirement of lifelong treatment, (iv) negative effect on immune cells, (v) drug adverse effects, (vi) drug-drug interactions, and (vii) drug resistance [7, 8].

Notably, if ART is started promptly in initial stages of infection, the latent reservoir is significantly reduced in size, but this reservoir is never completely eradicated [8, 9]. Regardless of the initiation of ART, the latent reservoir of HIV-1 infected cells is a momentous challenge in terms of limited immune response and potential for drug or vaccine treatment.

Lifelong treatment is necessary for continued viral load suppression and for the prevention of transmission to sexual partners, but there are some negative inevitable health consequences. Chronic inflammation and altered tissue architecture in lymph nodes and gastrointestinal tract occur on ART. This has significant repercussions on the functionality and survival of immune cells [8]. Additionally, long-term use is also associated with cardiovascular diseases, cancer, liver disease, long-term peripheral and central nervous complications, renal and metabolic disorders, and osteoporosis [4].

The adverse effects of ART drugs include insulin resistance, lipodystrophy, and dyslipidemia [8] and may lead to patient noncompliance. Additionally, given the barriers to accessibility, it may cause patients to be noncompliant and lead to issues with drug resistance [7].

Because of the difficulties of medication compliance, other ART drug deliveries are being tested in clinical trials, like long-acting injectable antiretrovirals, implantable devices, and vaginal rings [6]. Currently, there are long-acting injectable ART delivered by injections every two months being studied. The injection regimen increased patient adherence as compared to daily oral pills but has yet to be fully complete drug approval and implementation processes. Further research is aimed to extend the duration between injections to once yearly or twice a year [10].

3. Immune response

HIV-1 infection has a unique pathogenesis and degree of biological complexity unparalleled by other viruses. This biological complexity involves a wide range of immune cells and mechanisms, with the initial CD4 and CD8 T cell response, viral entry, and establishment of the latent reservoir reviewed in this section. Additionally, autologous neutralizing antibodies, broadly neutralizing antibodies (BnAbs), and extensive B cell dysfunction are important aspects to the immune response, but details are given in Sections 3.1–3.3.

In the initial stage after HIV-1 exposure but before infection, there are competing "HIV viral quasi-species" found in the donor secretions, vaginal mucosa of the recipient, and systemic component [11]. In the majority of infections (>75%), one variant referred to as the transmitted founder invades past the mucosal lumen to the stroma and travels to the local lymph nodes. (In 20% of infections, there are multiple variants involved in establishing systemic infection.) This transmitted founder has little genetic diversification. Here, at the lymph node, the CD4 T cells disperse the infection and exponential viral amplification, and systemic spread occurs [11].

Once systemic spread from the CD4 T cells is established, the CD8 T cells partially suppress the peak viral load after 30 days of infection. HIV rapidly evades the immune system and mutates to alter the epitope-human leukocyte antigen (HLA) binding. Over the next few weeks to months, the one viral variant becomes more like quasi-species version after exposure and diversifies its viral genome, due to the error-prone reverse transcriptase machinery adding different genetic codes to each infected cell [11, 12]. These viral variants may differ up to 40% of amino acids [13].

Though HIV-1 infection progresses in this pattern discussed in the last two paragraphs, understanding the specific mechanism of the virus entry is important aspect of the immune response. This mechanism begins with HIV envelope (Env) protein, gp120, binding CD4, and a cell surface protein receptor present on multiple types of immune cells. This gp120/CD4 interaction induces a conformational change, allowing gp120 to then bind to a coreceptor, CCR5, or CXCR4 [14–16]. Once coreceptor has been activated, gp41, another HIV-1 envelope protein, changes configuration to form a trimer-of-hairpins and fuses HIV-1 to the cell membrane, finalizing the entry [14].

The coreceptor of CCR5 or CXCR4 is of particular importance because this determines their susceptibility to infection. HIV-1 virus has two main "tropic" species, macrophage-tropic or M-tropic, and T-cell-tropic or T-tropic viruses [15, 16]. M-tropic strains are noted to be more preferentially transferred and utilize the CCR5 coreceptor, while T-tropic strains target the CXCR4 coreceptor [16]. Interestingly, there are some unique intermediate viruses referred to as "dual-tropic," meaning they can infect cells with either CXCR4 or CCR5 [15, 16]. Some drugs in ART regimens and in clinical trials involve mechanisms to antagonize the CCR5 or CXCR4 coreceptors and inhibit viral fusion inhibitors [14, 16]. Though drugs are effective, there are several associated problems, including multidrug resistance, adverse effects, and high cost [14]. Further research into understanding the mechanisms of HIV-1 entry and potential steps to terminate the infection may prove to be advantageous.

One of the major challenges of this virus is the early formation and somewhat permanent establishment of latent reservoir of infected CD4+ T cells. HIV-1 preferentially targets CD4+ T cells and then quickly reduces the memory CD4+ T cells in gut-associated lymphoid tissue in the first 4–10 days [17]. There are some studies that suggest that the initial prime target of the virus is epithelial Langerhans cells and dendritic cells during HIV-1 vaginal transmission, instead of the CD4 lymphocyte [11].

Regardless of the HIV-1 initial target, the impact of memory CD4+ T cell infection is significant because this establishes the latent reservoir and allows the virus to persist even with ART [18]. Utilizing nonhuman primate (NHP) simian immunodeficiency virus (SIV) studies, the latent reservoir is established within the first three days of infection, even before the virus is detectable [11].

Additionally, besides the establishment of the latent reservoir, there is another deleterious effect affecting CD4+ T lymphocytes in gastrointestinal mucosa. The Th17 subset of CD4+ T cells has an established role for maintaining mucosal integrity in the gastrointestinal tract. When these cells are infected, they no longer maintain a tight barrier or prevent gut bacterial products from invading the bowel wall [19]. This leads to microbial translocations, chronic immune activation, and shifts in microbiome [19–22].

3.1 Autologous neutralizing antibodies

Though the B cell response is less than optimal, as discussed in more detail in Section 6, the immune system is eventually able to produce autologous strain-specific neutralizing antibodies to the transmitted founder virus and the viral escape mutants. This occurs approximately three months to a year after infection [11]. These neutralizing antibodies are estimated to neutralize approximately 50% of the diverse strains of HIV-1 and only occur in half of patients with HIV-1 infections [12]. While these antibodies are ultimately overwhelmed by the continued viral evolution over the subsequent years, it is an important stepping stone for some individuals to progress generating broadly neutralizing antibodies [11], as reviewed in the next section.

3.2 Broadly neutralizing antibodies

Broadly neutralizing antibodies (BnAbs) are defined as antibodies with the ability to neutralize diverse isolates of HIV-1 and represent a potential avenue for prophylaxis and therapy [23]. This class of antibodies have high levels of mutations in rare, low-affinity naive B cell receptors [24], and these B-cell receptor (BCR) display atypical structural and binding characteristics that the immune system usually negatively selects against during B cell development [24, 25]. Some of the features of BnAbs are similar to those of autoreactive antibodies, indicating that they may evade immune tolerance mechanisms [26, 27]. Through the chronic viral replication, BnAbs are generated due to the extensive affinity maturation in germinal centers [27, 28].

BnAbs are an interesting and important phenomenon of the HIV-1 immune response, because only 10–20% of people with HIV-1 infections produce these antibodies after many years of infection [24, 25]. These individuals usually take

DOI: http://dx.doi.org/10.5772/intechopen.110222

approximately two to three years to produce BnAbs and are sometimes referred to as "elite neutralizers" [29]. While the BnAbs are able to neutralize many strains of HIV-1 variants, they ultimately are futile at controlling the host's infection because of years of sustained viremia [24].

Progress in the identification of BnAbs as well as the study of structural properties has significantly advanced the field, due to its application as potential therapy. To identify BnAbs, researchers can isolate HIV-1 Env-reactive memory B cells from multiple sources including antigen-specific B cell sorts, from plasma cell sorts, and from clonal memory B cell cultures [25].

The structural studies of BnAbs contribute to our understanding of the immune response to HIV-1 infection and its progression. Generally speaking, BnAbs perforate the glycan shield of the HIV Env trimer in five regions, likely dismantling Env function [23]. This massive glycan shield of the HIV Env trimer concealing the antigenic target is a major obstacle for the humoral immune system to produce antibodies that can neutralize HIV-1 variants [30].

Of these regions in the HIV Env trimer, the V2 apex is one of the most important because of its role in maintaining the metastability of the Env spike, which influences the CD4/CCR5 conformational changes. The CD4 binding site (CD4bs) is the primary receptor for HIV and exposes CCR5 after activation. Interfering with the V2 apex may prevent viral penetration. BnAbs targeting the V2 apex are considered a potent antibody but limited in breadth (less than 70%) and display incomplete neutralization (<100%) [23]. Further research is needed to elucidate this field.

3.3 B cell dysfunction

During acute HIV infection, multiple immune components lead to extreme B cell dysfunction. This is namely due to of polyclonal activation, hypergammaglobulinemia, nonspecific plasmablast surge, impaired memory and naïve B cells, and B cell exhaustion [28, 31].

Through polyclonal activation, B cells terminally differentiate into plasmablasts (cells rapidly produced in early antibody response to generate antibodies [32]) and plasma cells approximately a week after infection [11]. Polyclonal activation has been studied and shown to be elicited *via* multiple pathways directly from serum cytokines and indirectly from HIV Nef protein [28]. The polyclonal activation leads to a state of hypergammaglobulinemia [31].

Interestingly, though there is a state of hypergammaglobulinemia, the plasmablasts increase to comprise only up to 13% of circulating B cells. This response is not pathogen-specific since only 1.5% or less are HIV-specific and unable to neutralize the virus. In other viral infections like RSV and dengue virus, plasmablasts increase to comprise 30% of total lymphocytes, with the majority being pathogen specific [11, 33].

During HIV-1 infection, there is an increase in circulating antibodies and increase in activity of B cells, but a disruption in the microgenerative environment impacting memory and naïve B cell subsets [28]. This decline in circulating memory B cells is of particular importance and could be classified as a marker of disease progression, as they are linked to CD4+ T cell population numbers [34]. The low levels of memory B cells lead to the characteristic opportunistic infections, namely *Pneumocystis carinii* and *Cryptococcus neoformans* [34].

Among the B cell dysfunction category is B cell exhaustion, only recently described in 2008 in HIV. B cell exhaustion refers to the decreased ability to proliferate in response to *de novo* stimuli. HIV infections cause this exhaustion in a specific subset of B cells that are tissue-like memory B cells. This subset of cells has increased expression of multiple inhibitory receptors (CD22, CD72, and LAIR-1) as compared to normal resting memory and terminally differentiated B cells. Other unique inhibitory receptors to B cell exhaustion are also being investigated. Overall, B cell exhaustion is similar to CD4 and CD8 T cell exhaustion, but B cell exhaustion has proven to be more challenging to study since the experimental science is less direct as compared to T cell assays [31].

4. Variety of immune responses: progressors, immunological nonresponders, long-term nonprogressors, and elite controllers

One challenging aspect of HIV-1 infections is that there are a wide range of potential immune responses. There are four categories used to describe an individual's immune response to HIV-1: HIV-1 natural progressors, immunological nonresponders (INR), elite controllers, and long-term nonprogressors. Each category is described in detail below.

HIV-1 natural progressors refer to a typical response to an HIV-1 infection. The timeline for this response is not clearly defined but most likely reflects an intermediate progression where AIDS develops 3–10 years after seroconversion [35]. If these patients were to receive ART, the progression to AIDS would be less likely or potentially very slow and gradual.

Immunological nonresponders (INRs) have not been universally defined, but the most general accepted classification of INR is a patient who does not meet a specific CD4+ T cell count level or a specific percentage CD4+ T cell increase over baseline after a certain length of ART. The literature values for these specific levels and percentages vary widely with the CD4+ T cell count range of 200–500 and percentages from over 5–30%. The length of ART also changes from 6 to 144 months. Though the CD4+ T cell does not increase as it should, the viral load is suppressed. Given the inconsistent definition across studies, this subset is approximated at 10–40% of people with HIV-1 infections. This subset of people with HIV-1 infections is more likely to have morbidity and mortality from AIDs and non-AIDs conditions, because their immune systems are significantly dysfunctional [22].

Elite controllers and long-term nonprogressors represent those with immune responses that suppress HIV-1 viral load naturally without ART. These categories are differentiated by the degree of viral load suppression [36]. Elite controllers suppress the HIV-RNA values to less than 50 copies per milliliter, while long-term nonprogressors suppress the HIV-RNA to less than 5000 copies per milliliter [35].

Elite controllers are rare, estimated to be 0.1–1% of all people with HIV-1 infections but represent a functional cure. Understanding this population may lead to greater understanding of successful immune responses against HIV-1 [18]. As described in Loucif paper, elite controllers maintain suppressed viral loads most likely due to a combination of the following factors: high-quality and polyfunctional CD8+ T cells, memory B cell responses, preserved memory and pTfh CD4+ T cells, lack of natural killer (NK) activation, preserved plasmacytoid dendritic cell counts, and preservation of gut mucosal immunity [8].

The polyfunctionality of CD8+ T cells is a key differentiating factor between elite controllers and progressors. It has been found that the CD8+ T cells in elite controllers are able to degranulate properly and release perforin, granzyme, and cytokines (interferon-gamma, tumor necrosis factor-alpha, interleukin-2, and macrophage inflammatory protein-1beta) [36]. It is believed that the functional CD8+ T cell response is directly linked to disease progression [36].

Long-term progressors display similar characteristics as elite controllers, primarily the polyfunctional T cells [17]. They maintain high levels of CD4+ and CD8+ T cells without ART therapy. Approximately 5% of the total HIV population are long-term nonprogressors [21, 35].

4.1 Disadvantages for elite controllers and long-term nonprogressors

While the immune responses of elite controllers and long-term nonprogressors control the viral load in general, there are some downsides that these groups face. These patients can decline at any time, despite having long periods of naturally suppressed viral loads. While it is challenging to estimate the number of regressions in a small population of those with HIV-1 infections, it is believed that about 25% of elite controllers decline [37].

Researchers compared elite controllers who lost the ability to suppress the virus against the "persistent" elite controllers. The "persistent" elite controllers had low viral diversity, low HIV-DNA concentrations, overall lower inflammation, decreased immune activation, and proinflammatory cytokine concentrations. It was also found that the high Gag-specific T-cell polyfunctionality was no longer present in the individuals who lost viral control [37].

Though elite controllers represent a functional cure to HIV-1 infections, it is worth noting that this subset of patients is susceptible for hospitalizations of all causes (as compared to patients with HIV-1 infections on ART). The majority of these hospitalizations were due to cardiovascular and psychiatric diseases [36]. It is believed that a subset of elite controllers has this increased risk of hospitalizations and adverse effects because there is a persistent immune dysfunction driving the pathology [19].

5. Pediatric immune response

In addition to the various categories of immune responses, there is also a subset of young patients who acquire the infection perinatally or from breastfeeding from an HIV-1 positive mother. This occurs in part due to the large number of reproductive age women with limited access to ART and birth control in low-income countries. These patients have a different timeline than the standard adult, in terms of both immune response and overall disease course, and represent a significant public health crisis in low-income countries.

The timeline of the immune response in a pediatric patient begins with an infant with HIV-1 with high titers of passively transferred maternal neutralizing antibodies until three months of age. After this point, the neutralizing antibodies decrease but increase at 12 months, meaning that the infant is able to produce this antibody type. Some of these young patients then produce broadly neutralizing antibodies much earlier in infection, with diverse epitope specificities, and higher breadth and potency than that of adult patients [29]. One broadly neutralizing antibody (BF520.1) studied was noted to have limited somatic hypermutation and an absence of insertions and deletions, unlike the studies performed on adult antibodies. Given these core differences, the pediatric BnAbs are thought to be derived from different pathways than those produced in adults [29].

The overall disease course in a pediatric patient is unique because of a faster progression to AIDS [29] and a higher risk for neurocognitive deficits [38]. Without ART, children progress to AIDS within a year, as compared to adults taking a decade to progress [29]. These patients are more likely to have neurodevelopmental and neurocognitive disorders as compared to patients who acquired HIV-1 infection as adults. Given the rapid growth of nervous system in early infancy and childhood, understanding how HIV-1 infections impact childhood development is important. Pediatric patients are more likely to have physical brain damage from the inflammation and multinucleated giant cells in the cerebral cortex. The main manifestation of this impaired neurocognition is limitations in language function [38]. The studies of neurodevelopmental and neurocognitive effects on HIV-1 infection in pediatric patients are limited but represent a growing field of interest given the number of young patients in low-income countries [38].

Notably, 53% of untreated children with HIV-1 die by two years of age in sub-Saharan Africa. Before three years of age, this statistic changes to 75% children with HIV-1 dying [9]. Because of two modes of vertical transmission with HIV-1 infection occurring perinatally (in utero or intrapartum) and through breastfeeding, these groups of children have been assessed separately and identified that the children infected perinatally are at higher risk of death (60%) as compared to children infected through breast milk (36%) [9].

Given pediatric patients' immature immune systems and progression to AIDS, more research regarding acquisition prevention in these patients as well as funding is needed to combat this public health crisis in low-income countries.

6. Select clinical trials

Despite several decades of research, vaccine development, and clinical trials, currently, there is not any effective vaccine to prevent acquisition of HIV-1 infection. When HIV-1 was initially identified as the causative agent for AIDS in 1983–1984, researchers believed that a prophylactic vaccine would be generated within two years. This two-year estimate drastically underestimated the challenges and biological complexity of HIV-1 and illustrated the fact that HIV-1 is unlike any other viral disease that has a vaccine [4].

Though most clinical trials have found no efficacy, one clinical trial referred to as RV144 had unexpected success. This trial was controversial before it even began because it was believed to have a high likelihood of failure, given the early-phase clinical trials assessing the immunogens used in this trial. The initial data found the vaccine components were poorly immunogenic in isolation. Regardless of whether they were administered alone or in combination, there was only modest T-cell and humoral responses with no virus neutralization. However, the phase 3 trial proceeded in part to study a heterologous prime-boost strategy [39].

RV144 was the first trial that showed that any vaccine could induce protection against HIV-1 infection. Despite all odds, this vaccine had 60.5% efficacy in the first year [11] and decreased to 31.2% efficacy at 42-month post-vaccination [4]. Though the efficacy decreased significantly by three years, simulated studies believe that even if the vaccine had 50% efficacy for two years, it would have a significant impact in high prevalence areas [24].

As the only trial in humans with any efficacy, researchers had great interest in investigating what immune correlates were associated with protection against infection. The immune correlates for HIV-1 infection are unique inherently, given that the virus is never cleared naturally, but there is some type of protection against infection not yet understood. The immune correlates were identified as formation of non-neutralizing IgG against the V1/V2 region of HIV-1 Env, antibody-dependent cellular cytotoxicity in patients with low IgA, and Env-specific polyfunctional CD4+ T cells [4, 11, 24]. The mechanistic rationale behind how HIV-specific non-neutralizing antibodies protected against HIV-1 acquisition is not well understood and controversial [40].

In an effort to replicate RV144's success, a similar trial referred to as Uhambo or HVTN 702 was designed. There were clear differences between the two trials: RV144 had been conducted in Thailand in 2009, testing a recombinant Canarypox vector prime followed by two injections of a recombinant gp120 boost. HVTN 702 was conducted in South Africa with the same vector prime, similar protein boost but slightly different adjuvant, and different envelope sequences [24]. Investigators chose to change the envelope sequences to reflect the locally circulating HIV-1 variants in South Africa [41].

Ultimately, HVTN 702 was unsuccessful and terminated due to lack of efficacy in 2020 [41]. There was no significant production of the V2 loop antibodies deemed to be the critical immune correlate in RV144. This trial did result in high levels of binding antibodies, antibody-dependent cellular phagocytosis, and antibody-dependent cellular cytotoxicity activity, but overall no efficacy [42]. Perhaps, this lack of efficacy is due to the vast genetic diversity of the Sub-Saharan African with clade C, the difference in host genetic factors, or other indeterminate factors due to clinical trial differences as discussed by Gray et al. [41].

As previously discussed, the immune response to HIV-1 infection is intricate and complex with multiple stages of infection and potential responses.

7. Broadly neutralizing antibodies in passive immunization trials

Given the intricate method, the immune system forms BnAbs; immunization to induce BnAbs is proving to be exceedingly difficult. As previously reviewed in Section 5, the natural production is several years into infection and not clear why only a subset of people with HIV-1 infections generate these antibodies. The mechanism behind their evolution is also yet to be fully understood [25].

Before reviewing the vaccine strategies for BnAb induction, it is important to note that there have been clinical trials with passive immunization using VRC01, an IgG1 BnAb against the Env CD4 binding site. *In vitro* studies revealed that this BnAb has wide coverage against HIV-1 subtypes B and C. These trials were HVTN 704/HPTN 085 in the US, Peru, Brazil, and Switzerland and HVTN 703/HPTN 081 in South Africa, Zimbabwe, Malawi, Botswana, Kenya, Mozambique, and Tanzania, and ran from 2016 to 2018. The ultimate goal of these trials was to investigate if VRC01 is capable to preventing HIV-1 acquisition [43].

The results published in 2021 indicated that this BnAb was unable to prevent overall HIV-1 acquisition, but that in VRC01-sensitive HIV-1 isolates, BnAb prophylaxis was effective [43]. The VRC01-sensitive HIV-1 isolates were only ~30% of the strains in circulation [44]. These results suggest that the VRC01 suppressed early circulating strains, but the immune system eventually lost to evolving resistant viral variants. It is likely that a combination of BnAbs is necessary to prevent viral escape [43].

Additionally, "bispecific" and "trispecific" BnAbs were developed and tested in phase 1 clinical trials [43]. Bispecific or trispecific BnAbs have two or three different specificities in a single molecule. This unique class of BnAbs may lead to increased neutralization breadth and limit viral escape [45].

While studying how efficacious are BnAbs, it is important to assess their behavior *in vitro*, the administration of BnAbs *via* intravenous therapy is not a feasible drug delivery system for large populations. It may also have limited use if administered in combination of BnAbs in high-risk groups, but overall this is not feasible or sustainable method for HIV-1 prevention [44].

8. Promising vaccine strategies: general broadly neutralizing antibody vaccine

With the intricate and complex immune response to HIV-1 infection, it is no surprise that multiple vaccine strategies exist. Some promising vaccine strategies include SOSIP trimers, eOD-GT8 60mer and gene therapy, HIVcons immunogens, and mosaic immunogens. All these strategies are discussed in detail in Sections 9–12.

A feasible long-term solution or "the holy grail of HIV-1 vaccine development" is a vaccine that induces the production of BnAbs [25]. The current belief is that a prophylactic vaccine must induce BnAbs with a wide neutralization breath and/or HIV-1-specific antibodies to mediate antibody-dependent cellular cytotoxicity or other effector functions [39].

Multiple different concepts of BnAb-based vaccines are being investigated, including germline-targeted and lineage-based designs as well as SOSIP trimers detailed in Section 13 [27, 39]. In an effort to produce a germline-based vaccine, there are ongoing studies following HIV-1 isolates in people with active HIV-1 infections and BnAb [39]. A germline-targeted vaccine assesses an unmutated common ancestor of a BnAb, then approximates the critical BnAb precursor features, and designs the immunogens based on the structural and immunological information [27]. The goal of this germline approach is to preferentially activate B cells that are BnAb precursors and eventually produce memory B cells capable of BnAb production [46].

A lineage-based vaccine is focused on the immunological pathways forming antibody lineages that generate BnAb [27]. Regardless of which design is used, both germline-based and lineage-based designs use the "reverse vaccinology" to drive the production of the BnAb [44, 47].

These concepts are further complicated by the strategy of prime-boost vaccination models, where a priming vector is administered followed by a series of protein boosts. For this strategy to lead to BnAb production, the prime would be a precursor naive B cell, and the boosts would then select for neutralizing members of antibody lineage [27, 47].

9. Promising vaccine strategies: env glycoprotein trimers/SOSIP trimers

Another promising vaccine strategy that may result in BnAb production is immunogens based off the recombinant stabilized HIV envelope (Env) glycoprotein trimers, specifically the SOSIP trimers [47, 48]. SOSIP trimers are proteins that adopt a native-like confirmation and strongly approximate the features of HIV-1 virion [46]. Currently, the SOSIP trimers induce neutralizing antibodies against "relatively neutralization resistant (tier 2) autologous viruses" but there are multiple studies suggesting their neutralization breadth [39, 47, 49], as well as improvements in their production [39, 50], stabilization [46], and immunogenicity as future vaccine candidates [51].

To study the neutralization breadth, one study was completed through computational modeling that assessed different combination and sequential vaccine series and then administered to groups of rabbits. While this experiment did not induce BnAb formation, it proved that the neutralizing antibody response can be cross-boosted when Env trimers from different clades were administered sequentially [49]. Further research is needed to identify a SOSIP trimer that induces BnAb production.

One study determined the adjuvants compatible with SOSIP trimers, since the adjuvant is a key component to boost the immune response against the vaccine antigens. This study found that different adjuvant classes did not interfere with the integrity of the SOSIP trimer, with the exception of aluminum sulfate formulations [51].

Additionally, to assess how feasible and scalable SOSIP production was, Dey et al. completed a trial that resulted 3.52 grams of fully purified trimers. The overall production was performed while abiding by current Good Manufacturing Practice guidelines. The quality of SOSIP composition was compared to samples generated from a research laboratory and reported to be the same in terms of antigenicity, disulfide bond patterns, and glycan composition. This trial also helped assess how stable the trimers were at different temperatures and storage conditions [50].

10. Promising vaccine strategies: eOD-GT8 60mer and gene therapy

Additional BnAb-based vaccines currently being assessed are eOD-GT8 60 mer (protein nanoparticle derived from the VRC01 CD4-binding site) and gene therapy through a recombinant adeno-associated virus-mediated (rAAV) construct containing HIV BnAb genes [44].

This eOD-GT8 60mer delivered with AS01B adjuvant was tested in 2018 in an early clinical trial. This nanoparticle was originally studied in genetically modified knock-in mice and animal studies and designed to activate B-cell precursors to form BnAb. The results showed that 97% subjects produced precursor VRC01 IgG B cells, according to the preliminary data in early 2021 at the 2021 R4P conference. This is an important finding, because it provides some evidence that stimulation of rare B-cell precursors is possible. Further study is needed to validate this approach and determine that it works consistently [44].

A gene therapy approach to delivery BnAb genes utilizes the rAAV vector and selects genes like VRC01, b12, 4E10, 2F5, 2G12, 3BNC117, 10–1074, and 10E8. This strategy was assessed in rhesus macaques and led to undetectable viremia for three years. While this was incredibly positive, the animals developed antidrug antibody (ADA) responses. Further testing with nonhuman primates (NHP) yielded similar results, with a decline of a BnAb corresponding to ADA response. In the NHP study, these animals did have mucosal protection against the simian version of HIV but for a shorter duration (less than a month). This may have been due to only receiving one BnAb gene in the rAAV vector. The ADA response remains the main hurdle in this approach, provided that the BnAb gene combinations are validated [44].

These initial data from both the eOD-GT8 60 mer and rAAV vector with BnAb gene immunization strategies are encouraging. More research and trials are necessary to fully understand their potential to generate a prophylactic HIV-1 immunization.

11. Promising vaccine strategies: HIVcons immunogen-based vaccine

A promising concept that may result in a prophylactic HIV-1 vaccine is one with a HIVcons immunogen basis. This design is a 778 amino acid insert created analyzing all clades through bioinformatics. The insert specifically focused on sequence conservation and did not use epitopes as a "string of beads" [13]. String of beads epitope design refers to a technique where short sequences of amino acids or spacers are added between epitopes in an attempt to allow the epitopes to be cleaved correctly [52]. Researchers creating HIVcons immunogen elected to not use the string of beads construct, because it could create irrelevant epitopes not present on HIV-1-infected cells. Additionally, this construct could skew the design toward frequent Caucasian HLA alleles [13].

While the HIVcons insert was successfully tested in multiple early clinical trials in two different vectors in both healthy volunteers and patients with HIV-1 infections, researchers had to redesign to create a "second-generation conserved mosaic tHIVconsvX" vaccine, utilizing the CdAdOx1 vector from simian adenovirus Y25. This was due to licensure issues with the previous vectors, but believed to be advantageous overall since the second-generation design was improved, as discussed by Tomas Hanke [13].

12. Promising vaccine strategies: mosaic immunogens

Mosaic immunogens is another vaccine design that is showing promising results in clinical trials. These immunogens are bioinformatically optimized bivalent antigens derived from different clades of HIV-1 group M strain [53]. These sequences were then placed into MVA or Ad26 vectors, followed by a clade C gp140 protein boost. This was evaluated in the APPROACH study in East Africa, South Africa, Thailand, and the USA. The results indicated that the Ad26/Ad26+ high dose gp140 vaccinations produced Env-specific antibody responses and T cell responses in the vast majority of recipients (100%, 83% for each of these responses) [54].

Building off the APPROACH study data as well as TRAVERSE and ASCENT studies, two studies, Imbokodo (HVTN 705) and Mosaico (HVTN 706), were designed. Imbokodo was phase 2b efficacy clinical trial for women in Southern Africa evaluating a heterologous prime/boost of Ad26.Mos4.HIV and adjuvanted aluminum phosphate Clade C gp140 [6, 44]. In total, it has four mosaic antigens within the vector aimed at inducing an immune response. Preliminary results of Imbokodo showed no significant efficacy (25.2% efficacious with 95% confidence interval between −10.5% and 49.3%), but these data are yet to be published in a peer-reviewed journal [55].

Though Imbokodo was completed, Mosaico, the similar trial, is not yet finished and has some differences in vaccine components and patient population [53]. This vaccine has gp140 immunogens from different HIV strains, rather than clade C specific as in Imbokodo [6, 44]. Mosaico enrolled gay men and other men who have sex with men as well as transgender women. This study is still ongoing in hopes that the slight differences between immunogens and transmission mode will result in some vaccine efficacy [55, 56].

DOI: http://dx.doi.org/10.5772/intechopen.110222

13. Conclusion

There is great need for a prophylactic HIV-1 vaccination to end the HIV/AIDS epidemic. HIV-1 immunology and vaccine development remain a uniquely challenging field. This is for multiple factors: complex immune mechanisms, rapid mutation of the viral variants, establishment of the latent reservoir early in infection, no predictive animal models or natural immune correlates, and only one efficacious clinical vaccine trial with RV144 in 2009 with numerous failed trials. Despite these factors, the field has made great strides, particularly through the study of categorization of immune responses and characterization of broadly neutralizing antibodies (BnAbs). This progress has led to multiple promising vaccine candidates in various stages of development, namely, SOSIP trimers, eOD-GT8 60mer and gene therapy, HIVcons immunogens, and mosaic immunogens.

Conflict of interest

The authors declare no conflict of interest.

Author details

Anna Hargrave[1], Abu Salim Mustafa[2], Asma Hanif[3], Javed Tunio[4] and Shumaila Nida Muhammad Hanif[5]*

1 Kentucky College of Osteopathic Medicine, University of Pikeville, Pikeville, USA

2 Faculty of Medicine, Department of Microbiology, Kuwait University, Kuwait City, Kuwait

3 Department of Restorative Sciences, College of Dentistry, Kuwait University, Kuwait

4 Department of Internal Medicine, Carver College of Medicine, University of Iowa, USA

5 Department of Biomedical Sciences, Kentucky College of Osteopathic Medicine, University of Pikeville, Pikeville, USA

*Address all correspondence to: dr.shumaila_nida@yahoo.com

References

[1] UNAIDS. Global HIV & AIDS statistics – Fact sheet. [Internet]. 2022. Available from: https://www.unaids.org/en/resources/fact-sheet. [Accessed October 10, 2022]

[2] Kharsany ABM, Karim Q. HIV infection and AIDS in sub-Saharan Africa: Current status, challenges and opportunities. The Open AIDS Journal. 2016;**10**:34-38. DOI: 10.2174/1874613601610010034

[3] Piot P, Bartos M, Ghys PD, et al. The global impact of HIV/AIDS. Nature. 2001;**410**:968-973. DOI: 10.1038/35073639

[4] Nguni T, Chasara C, Ndhlovu ZM. Major scientific hurdles in HIV vaccine development: Historical perspective and future directions. Frontiers in Immunology. 2020;**11**:590780. DOI: 10.3389/fimmu.2020.590780

[5] Bekker LG, Tatoud R, Dabis F, et al. The complex challenges of HIV vaccine development require renewed and expanded global commitment. Lancet. 2020;**395**:384-388. DOI: 10.1016/S0140-6736(19)32682-0

[6] Stephenson KE, Wagh K, Korber B, et al. Vaccines and broadly neutralizing antibodies for HIV-1 prevention. Annual Review of Immunology. 2020;**38**:673-703. DOI: 10.1146/annurev-immunol-080219-023629

[7] Kalidasan V, Das KT. Lessons learned from failures and success stories of HIV breakthroughs: Are we getting closer to an HIV cure? Frontiers in Microbiology. 2020;**11**:46. DOI: 10.3389/fmicb.2020.00046

[8] Loucif H, Gouard S, Dagenais-Lussier X, et al. Deciphering natural control of HIV-1: A valuable strategy to achieve antiretroviral therapy termination. Cytokine & Growth Factor Reviews. 2018;**40**:90-98. DOI: 10.1016/j.cytogfr.2018.03.010

[9] Luzuriaga K, Mofenson LM. Challenges in the elimination of Pediatric HIV-1 infection. New England Journal of Medicine. 2016;**374**:761-770. DOI: 10.1056/NEJMra1505256

[10] Fauci AS, Dieffenbach CW, Dabis F. Fitting a vaccine into the HIV prevention landscape. Journal of the International AIDS Society. 2021;**24**:e25792. DOI: 10.1002/jia2.25792

[11] Gao Y, McKay PF, Mann JFS. Advances in HIV-1 vaccine development. Viruses. 2018;**10**:167. DOI: 10.3390/v10040167

[12] Kwong PD, Mascola JR. HIV-1 vaccines based on antibody identification, B cell ontogeny, and epitope structure. Immunity. 2018;**48**:855-871. DOI: 10.1016/j.immuni.2018.04.029

[13] Hanke T. Aiming for protective T-cell responses: A focus on the first generation conserved-region HIVconsv vaccines in preventive and therapeutic clinical trials. Expert Reviews of Vaccines. 2019;**18**:1029-1041. DOI: 10.1080/14760584.2019.1675518

[14] Yamada Y, Ochiai C, Yoshimura K, et al. CD4 mimics targeting the mechanism of HIV entry. Bioorganic and Medicinal Chemistry Letters. 2010;**20**(1):354-358. DOI: 10.1016/j.bmcl.2009.10.098

[15] Doms RW. Chemokine receptors and HIV entry. AIDS.

2001;**15**(Suppl 1):S34-S35. DOI: 10.1097/00002030-200102001-00051

[16] Esté JA, Telenti A. HIV entry inhibitors. Lancet. 2007;**370**(9581):81-88. DOI: 10.1016/S0140-6736(07)61052-6

[17] Barouch DH. Challenges in the development of an HIV-1 vaccine. Nature. 2008;**455**:613-619. DOI: 10.1038/nature07352

[18] Chun TW, Carruth L, Finzi D, et al. Quantification of latent tissue reservoirs and total body viral load in HIV-1 infection. Nature. 1997;**387**:183-188. DOI: 10.1038/387183a0

[19] Cai CW, Sereti I. Residual immune dysfunction under antiretroviral therapy. Seminars in Immunology. 2021;**51**:101471. DOI: 10.1016/j.smim.2021.101471

[20] Barouch DH, Korber B. HIV-1 vaccine development after STEP. Annual Review of Medicine. 2010;**61**:153-167. DOI: 10.1146/annurev.med.042508.093728

[21] Shi Y, Su J, Chen R, et al. The role of innate immunity in natural elite controllers of HIV-1 infection. Frontiers in Immunology. 2022;**13**:780922. DOI: 10.3389/fimmu.2022.780922

[22] Yang X, Su B, Zhang X, et al. Incomplete immune reconstitution in HIV/AIDS patients on antiretroviral therapy: Challenges of immunological non-responders. Journal of Leukocyte Biology. 2020;**107**:597-612. DOI: 10.1002/JLB.4MR1019-189R

[23] Sok D, Burton DR, Recent progress in broadly neutralizing antibodies to HIV [published correction appears in Nature Immunology]. Nature Immunology. 2019;**19**:1179-1188. DOI: 10.1038/s41590-018-0235-7

[24] Lopez Angel CJ, Tomaras GD. Bringing the path toward an HIV-1 vaccine into focus. PLoS Pathogens. 2020;**16**:e1008663. DOI: 10.1371/journal.ppat.1008663

[25] Haynes BF, McElrath MJ. Progress in HIV-1 vaccine development. Current Opinion in HIV and AIDS. 2013;**8**:326-332. DOI: 10.1097/COH.0b013e328361d178

[26] Williams WB, Wiehe K, Saunders KO, et al. Strategies for induction of HIV-1 envelope-reactive broadly neutralizing antibodies. Journal of International AIDS Society. 2021;**24**:e25831. DOI: 10.1002/jia2.25831

[27] Haynes BF, Burton DR, Mascola JR. Multiple roles for HIV broadly neutralizing antibodies. Science Translational Medicine. 2019;**11**:eaaz2686. DOI: 10.1126/scitranslmed.aaz2686

[28] Shen X, Tomaras GD. Alterations of the B-cell Response by HIV-1 Replication [published correction appears in Current HIV/AIDS Reports]. Current HIV/AIDS Reports. 2011;**2011**(8):23-30. DOI: 10.1007/s11904-010-0064-2

[29] Kumar S, Panda H, Makhdoomi MA, et al. An HIV-1 broadly neutralizing antibody from a clade C-infected Pediatric elite neutralizer potently neutralizes the contemporaneous and autologous evolving viruses. Journal of Virology. 2019;**93**:e01495-e01418. DOI: 10.1128/JVI.01495-18

[30] Lee JH, Crotty S. HIV Vaccinology: 2021 Update. Seminars in Immunology. 2021;**51**:101470. DOI: 10.1016/j.smim.2021.101470

[31] Moir S, Fauci AS. B-cell exhaustion in HIV infection: The role of immune activation. Current Opinion HIV and

AIDS. 2014;**9**:472-477. DOI: 10.1097/COH.0000000000000092

[32] Nutt SL, Hodgkin PD, Tarlinton DM, et al. The generation of antibody-secreting plasma cells. Nature Reviews Immunology. 2015;**15**:160-171. DOI: 10.1038/nri3795

[33] Wrammert J, Onlamoon N, Akondy RS, et al. Rapid and massive virus-specific Plasmablast responses during acute dengue virus infection in humans. Journal of Virology. 2012;**86**:2911-2918. DOI: 10.1128/JVI.06075-11

[34] Cagigi A, Nilsson A, De Milito A, et al. B cell immunopathology during HIV-1 infection: Lessons to learn for HIV-1 vaccine design. Vaccine. 2008;**26**:3016-3025. DOI: 10.1016/j.vaccine.2007.11.063

[35] Kumar P. Long term non-Progressor (LTNP) HIV infection. Indian Journal of Medical Research. 2013;**138**:291-293

[36] Gebara NY, El Kamari V, Rizk N. HIV-1 elite controllers: An Immunovirological review and clinical perspectives. Journal of Virus Eradication. 2019;**5**:163-166

[37] Tarancon-Diez L, Rodríguez-Gallego E, Rull A, et al. Immunometabolism is a key factor for the persistent spontaneous elite control of HIV-1 infection. eBioMedicine. 2019;**42**:86-96. DOI: 10.1016/j.ebiom.2019.03.004

[38] Williams ME, Janse Van Rensburg A, Loots DT, et al. Immune dysregulation is associated with neurodevelopment and neurocognitive performance in HIV Pediatric populations-a scoping review. Viruses. 2021;**13**:2543. DOI: 10.3390/v13122543

[39] Jones LD, Moody MA, Thompson AB. Innovations in HIV-1 vaccine design. Clinical Therapeutics. 2020;**42**:499-514. DOI: 10.1016/j.clinthera.2020.01.009

[40] Feinberg MB. Uhambo - twists and turns on the journey to an efficacious HIV-1 vaccine. New England Journal of Medicine. 2021;**384**:1157-1159. DOI: 10.1056/NEJMe2102358

[41] Gray GE, Bekker LG, Laher F, et al. Vaccine efficacy of ALVAC-HIV and bivalent subtype C gp120-MF59 in adults. New England Journal of Medicine. 2021;**384**:1089-1100. DOI: 10.1056/NEJMoa2031499

[42] Gray GE, Corey L. The path to find an HIV vaccine. Journal of International AIDS Society. 2021;**24**:e25749. DOI: 10.1002/jia2.25749

[43] Corey L, Gilbert PB, Juraska M, et al. Two randomized trials of neutralizing antibodies to prevent HIV-1 acquisition. New England Journal of Medicine. 2021;**384**:1003-1014. DOI: 10.1056/NEJMoa2031738

[44] Sobia P, Archary D. Preventive HIV vaccines-leveraging on lessons from the past to pave the way forward. Vaccine. 2021;**9**:1001. DOI: 10.3390/vaccines9091001

[45] Pegu A, Xu L, DeMouth ME, et al. Potent anti-viral activity of a Trispecific HIV neutralizing antibody in SHIV-infected monkeys. Cell Reports. 2022;**38**:110199. DOI: 10.1016/j.celrep.2021.110199

[46] Steichen JM, Kulp DW, Tokatlian T, et al. HIV vaccine design to target germline precursors of glycan-dependent broadly neutralizing antibodies. Immunity. 2016;**45**:483-496. DOI: 10.1016/j.immuni.2016.08.016

[47] Burton DR. Advancing an HIV vaccine; advancing vaccinology. Nature Reviews Immunology. 2019;**19**:77-78. DOI: 10.1038/s41577-018-0103-6

[48] Trovato M, D'Apice L, Prisco A, et al. HIV vaccination: A roadmap among advancements and concerns. International Journal of Molecular Sciences. 2018;**19**:1241. DOI: 10.3390/ijms19041241

[49] de la Torrents Peña A, de Taeye SW, Sliepen K, et al. Immunogenicity in rabbits of HIV-1 SOSIP trimers from Clades A, B, and C, given individually, sequentially, or in combination. Journal of Virology. 2018;**92**:e01957-e01917. DOI: 10.1128/JVI.01957-17

[50] Dey AK, Cupo A, Ozorowski G, et al. cGMP production and analysis of BG505 SOSIP.664, an extensively glycosylated, trimeric HIV-1 envelope glycoprotein vaccine candidate. Biotechnology. Bioengineering. 2018;**115**:885-899. DOI: 10.1002/bit.26498

[51] Ozorowski G, Cupo A, Golabek M, et al. Effects of adjuvants on HIV-1 envelope glycoprotein SOSIP trimers In vitro. Journal of Virology. 2018;**92**:e00381-e00318. DOI: 10.1128/JVI.00381-18

[52] Dorigatti E, Schubert B. Joint epitope selection and spacer Design for String-of-Beads Vaccines. Bioinformatics. 2020;**36**:i643-i650. DOI: 10.1093/bioinformatics/btaa790

[53] Barouch DH, Tomaka FL, Wegmann F, et al. Evaluation of a mosaic HIV-1 vaccine in a multicentre, randomised, double-blind, placebo-controlled, phase 1/2a clinical trial (APPROACH) and in Rhesus monkeys (NHP 13-19). Lancet. 2018;**392**:232-243. DOI: 10.1016/S0140-6736(18)31364-3

[54] Pitisuttithum P, Marovich MA. Prophylactic HIV vaccine: Vaccine regimens in clinical trials and potential challenges. Expert Review of Vaccines. 2020;**19**:133-142. DOI: 10.1080/14760584.2020.1718497

[55] National Institutes of Health. HIV Vaccine Candidate Does Not Sufficiently Protect Women Against HIV Infection. [Internet]. 2022. Available from: https://www.nih.gov/news-events/news-releases/hiv-vaccine-candidate-does-not-sufficiently-protect-women-against-hiv-infection. [Accessed November 11, 2022]

[56] Hannah S, Chinyenze K, Shattock R, et al. HIV vaccines in 2022: Where to from here? Journal of the International AIDS Society. 2022;**25**:e25923. DOI: 10.1002/jia2.25923

Chapter 3

Perspective Chapter: Iron Chelation Inhibits Reduced Glutathione (GSH) as a Prooxidant in Iron-Mediated Hemoglobin Oxidation in Purified Hemoglobin

Ibrahim Mustafa

Abstract

As a trace element, iron is required by all living. Although this crucial metal is required, maintaining its biological equilibrium in an organism is far more important than any other trace element. Excess iron plays a vital role in the generation of harmful oxygen radicals due to its catalysis of one electron redox chemistry. In disorders such as thalassemia and sickle cell anemia, this is clearly visible. In vitro experiments were carried out using pure hemoglobin (HbA) exposed to ferric (Fe^{3+}) iron. The addition of Fe^{3+} (0–250 μM) caused spectrophotometric alterations in the absorption spectra (500–700 nm) of (40 μM HbA; pH 7.4). There was no HbA oxidation in the absence of Fe^{3+}. Similarly, unlike hemolysates, the mere addition of Fe^{3+} to HbA exhibited negligible oxidative consequences. However, the addition of glutathione (GSH) and Fe^{3+} caused significant oxidation. The iron chelators (DFO desferrioxamine or Deferiprone L1) suppressed Fe^{3+}-mediated HbA oxidation in a dose-dependent manner. The findings of this study have important significance for damage mechanisms in disorders like as thalassemia and sickle cell anemia. In addition, our findings suggest that chelating bioreactive iron within aberrant erythrocytes might be a potential therapy strategy.

Keywords: iron, deferoxamine, glutathione, hemoglobin, oxidation

1. Introduction

Iron is a vital trace element for all living cells. Despite the fact, this key metal is essential, and maintaining its biological balance in an organism is far more important than any other trace element except copper [1]. Excess iron, due to its catalysis of one electron redox chemistry, plays a key role in the formation of toxic oxygen radicals. Indeed, this is readily observed in diseases such as thalassemia and sickle cell anemia. This potentially hazardous combination of oxygen and iron within the erythrocyte is kept in check by several endogenous mechanisms. Intra-erythrocytic free iron can be a potential hazard to form free radicals in the presence of reduced glutathione. Free radicals can have an adverse effect on hemoglobin by oxidative damage [2]. Glutathione (GSH) is thought to be a prooxidant in iron-mediated hemoglobin oxidation, which can be prevented by iron chelation. Researchers picked two *in vitro* hemoglobin models to test this theory. RBC hemolysate, which is essentially distilled water lysed RBC, and crude pure hemoglobin (Hb-A) eluted off a Sephadex desalting column.

2. Experimental methods used in this study

A random fresh blood sample collected in EDTA tube was obtained from Hamad Medical Corporation (HMC), Doha, Qatar. The blood samples were obtained from HMC blood donor center in Doha, Qatar. Institutional review board (IRB) approval was obtained from HMC for using the donor's blood for the purpose of research. This research has two model samples (the hemolysate and the purified hemoglobin). The principal materials were: iron (III) solution, iron chelators (deferoxamine and deferiprone), and glutathione. Dry-ice acetone bath, hot water bath, centrifuge, UV-VIS spectrophotometer (absorption spectroscope), and disposable PD-10 desalting columns kit.

2.1 Preparation of hemolysate

Plasma was isolated from the entire blood sample for the hemolysate sample preparation by centrifugation at 2000 rpm. RBCs were washed three times with normal saline, 5 ml of distilled water was put into a fresh test tube, and 20 μl of packed RBCs were added to the 5 ml distilled water tube, which was gently mixed. Distilled water promotes RBC hemolysis, which results in the creation of hemolysate.

2.2 Preparation of crude purified hemoglobin

For purified hemoglobin preparation, pRBCs tube was placed in a dry-ice acetone bath for few seconds until the PRBCs freeze and visibly seen as solid. The tube immediately was removed from the dry-ice acetone bath to thaw in the hot water bath. Sudden cooling and thawing three times cause proteins of erythrocyte's membrane to denature and eventually lead to cell lysis. After the pRBCs are lysed, they will be placed in the PD 10 desalting column. The PD-10 desalting columns are intended for the fast removal of proteins and other big macromolecules from samples [3]. The PD-10 desalting columns are used to capture large-molecular weight chemicals and proteins, particularly glutathione (307.32 g/mol), superoxide dismutase, and catalase. Eluted purified hemoglobin (64,458 g/mol) from the PD-column will be collected in another tube. The working solution of HbA is made with 40μM HbA at pH 7.4 in water.

2.3 Hemoglobin oxidation studies

Prepare stocks of iron, glutathione, deferoxamine, and deferiprone solution and dilute them with distilled water to reach the exact quantities and concentrations required for the experiment. The absorbance spectra of hemoglobin were measured using a UV-VIS spectrophotometer, either in hemolysate or pure form.

"There is no discernible difference in the visible area between the absorption spectra obtained from hemolysate and pure hemoglobin," writes Horecker (1943). To view the distinctive spectrum activity of oxy-Hb, the oxy-Hb must be scanned in

(500–700 nm) wavelength before adding the iron or chelators to analyze their influence on oxyhemoglobin spectra, oxy-Hb has two distinct peaks at 541 and 577 nm. The oxidation of hemoglobin was evaluated by spectrophotometric analysis (500–700 nm); the concentrations of oxy-, met-, and hemichrome hemoglobin were estimated using the Winterbourn technique.

After assessing the effect of iron on oxy-Hb in both models, hemolysate and pure Hb, the iron concentration that causes the most oxidative damage will be chosen for the next tests. Because hemolysate contains its own glutathione, we shall solely evaluate the impact of reduced glutathione on purified Hb. The prepared glutathione solution, as well as precise iron concentrations, will be added. The impact of the iron chelator will then be investigated.

Microsoft Excel was used for statistical analysis. The concentration of oxyhemoglobin was determined using the Winterbourn technique from the absorption spectra. Experiments were carried out in triplicate at each time point, and the scanning data displayed are representative of the outcomes of the experiments.

3. Results and data analysis

Figures 1 and **2** show the findings of spectrum measurements in iron-mediated oxyhemoglobin oxidation on hemolysate with regard to time in minutes. The UV-visible spectra in **Figure 1A** and **B** indicate the effect of dose-dependent iron-mediated oxidative damage on oxyhemoglobin in hemolysate during the first 10 minutes of iron addition. **Figure 2** is visibly expressing the effect of iron by causing oxidative damage to the oxyhemoglobin content of hemolysate. The hemolysate is losing the normal red pigmentation of the hemoglobin into brown color due to iron effect with different concentrations concerning time (5 minutes).

The results of the spectral measurements in iron-mediated oxyhemoglobin oxidation on purified hemoglobin only with respect to the time that was measured

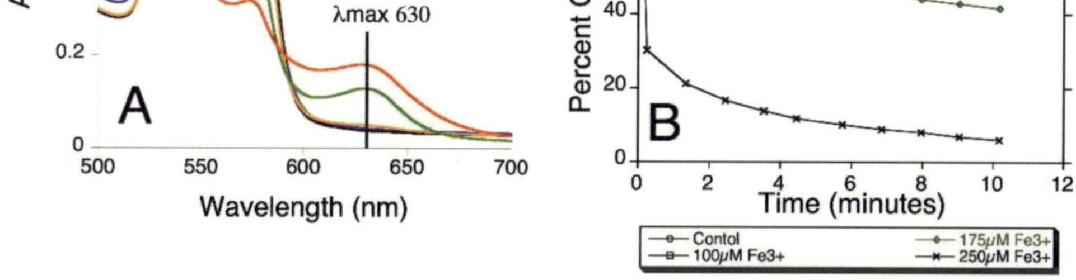

Figure 1.
UV-visible spectra of dose-dependent iron-mediated oxidative damage on oxyhemoglobin in hemolysate after 10 minutes of iron addition are shown in A. The % oxyhemoglobin derived from the spectral shift with dose-dependent iron-induced hemoglobin oxidation in hemolysate is shown in B.

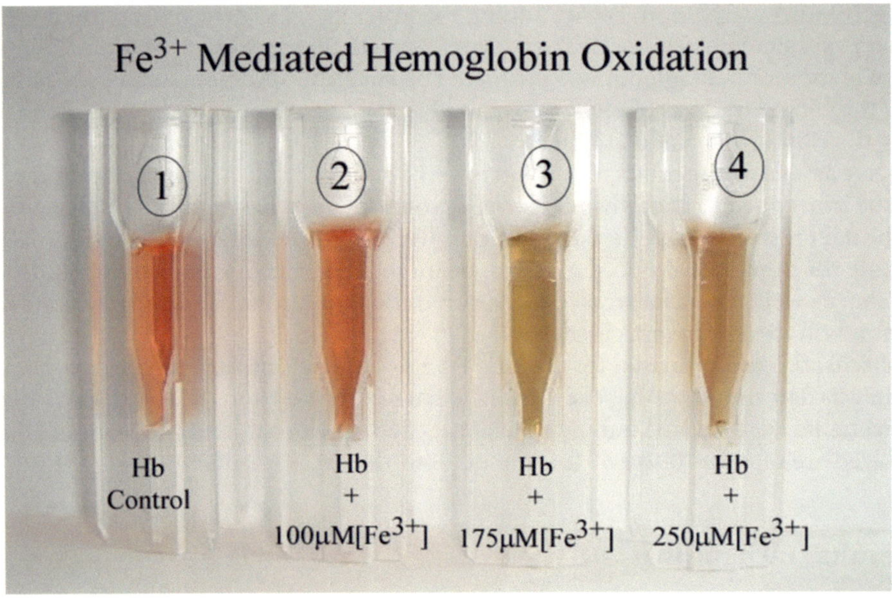

Figure 2.
Figure demonstrating the effect of iron on hemolysate oxyhemoglobin concentration, which causes oxidative damage. Because of the iron activity in variable levels throughout time, the hemolysate loses its normal red color and turns brown (10 minutes). (1) represents the control, (2) represents 100 µM Fe^{3+}, (3) represents 175 µM Fe^{3+}, and (4) represents 250 µM.

in minutes are shown in **Figure 3**. The UV-visible spectra shows the effect of dose-dependent iron-mediated oxidative damage on oxyhemoglobin in purified hemoglobin during the first minute of iron addition, where there was no visible change. **Figure 4** visibly expresses the effect of iron in causing oxidative damage to the oxyhemoglobin content of purified hemoglobin. There is no visible change in the normal red pigmentation of the hemoglobin of different iron concentrations concerning time (5 minutes). **Figure 5** is a graph showing the binding of iron chelators with the iron. As the iron

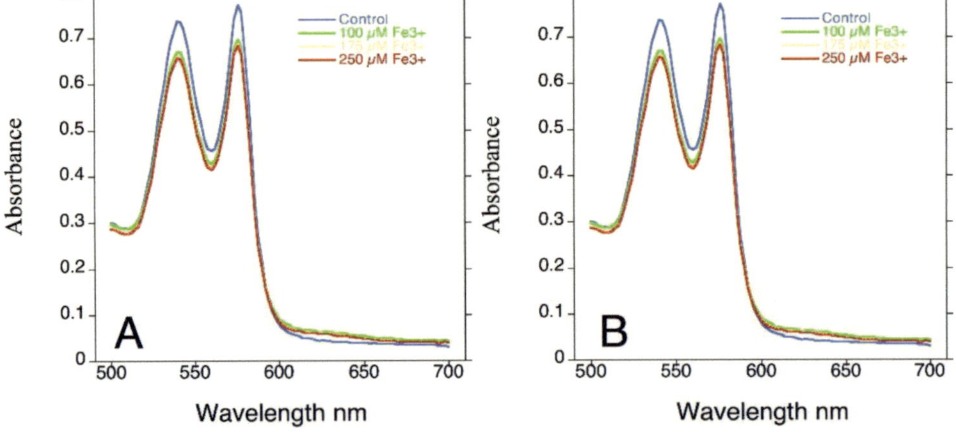

Figure 3.
A: UV-visible spectra showing the effect of dose-dependent iron-mediated oxidative damage on oxyhemoglobin in purified hemoglobin after 5 minutes of iron addition. B shows there is no change in spectra after 10 minutes.

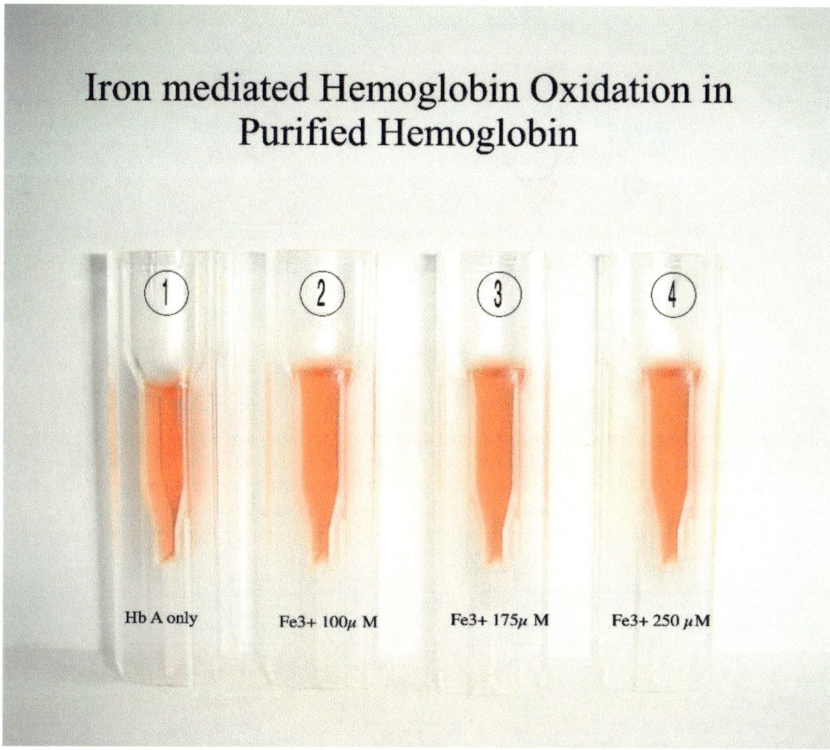

Figure 4.
Figure visibly expressing the effect of iron in causing oxidative damage to the oxyhemoglobin content of purified hemoglobin. There is no visible change in the normal red pigmentation of the hemoglobin different iron concentrations with respect to time (5 minutes). (1): the control, (2): 100 μM Fe^{3+}, (3): 175 μM Fe^{3+}, and (4): 250 μM Fe^{3+}.

chelators are added to the iron in the cuvette, the color changed. Also, the absorbance is changed indicating a reaction (peaks) between iron and the chelating agent. As shown, the red and black lines, where the iron is bound with chelators are showing noticeable peaks that are indicative of a binding reaction. The maximum absorbance of L1-Fe and DFO-Fe complexes is achieved at 450 to 470 nm and 430 to 460 nm, respectively [4].

Figure 6 depicts iron-mediated hemoglobin oxidation and the inhibitory impact of DFO and L1 in hemolysate. In A, the addition of 250 μM results in a significant drop in the percentage of oxyhemoglobin, and when the iron chelator DFO is applied, the percentage of oxyhemoglobin is recovered to above 90%. Similarly, in B, a bidentate iron chelator L1 has a similar response, inhibiting iron-mediated hemoglobin oxidation [5].

Figure 7 depicts the iron-mediated oxidation of pure hemoglobin, with no substantial loss of % oxyhemoglobin in purified hemoglobin solution as compared to hemolysate. However, the mixing of iron and glutathione in pure hemoglobin produced results comparable to hemolysate. Iron chelator DFO inhibits iron and glutathione-mediated hemoglobin oxidation in A, whereas iron chelator L1 inhibits comparable findings in B.

4. Discussions

Excess bio-reactive iron contributes to the formation of free radicals, which can lead to oxidative damage to the hemoglobin in our red blood cells. This oxidative

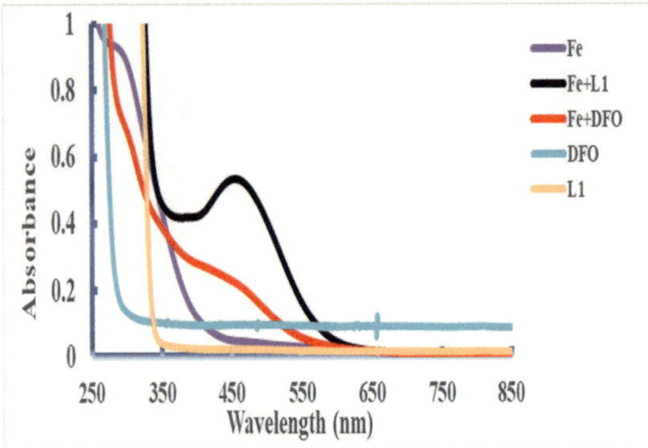

Figure 5.
The graph depicts the binding of iron chelators to iron; as the iron chelators are added to the iron in the cuvette, the color changes, suggesting a reaction (peaks) between iron and the chelating agent. The red and black lines, where the iron is bonded with chelators, display prominent peaks, indicating a binding response. L1-Fe and DFO-Fe complexes had maximal absorbance at 450 to 470 nm and 430 to 460 nm, respectively [1].

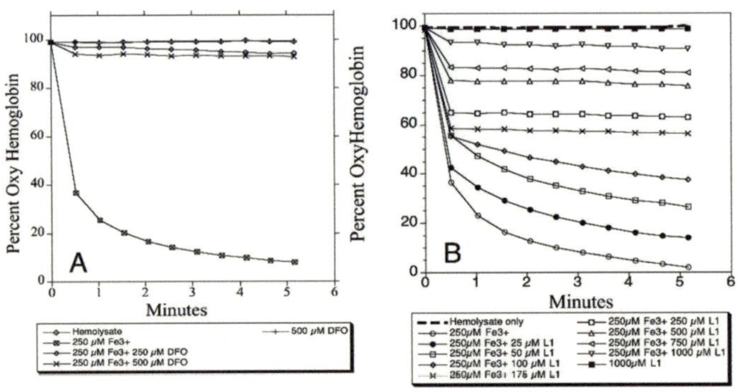

Figure 6.
A: Fe^{3+}-mediated hemoglobin oxidation in hemolysate. Shown is the oxyhemoglobin concentration following the addition of exogenous 250 μM Fe^{3+} and DFO in hemolysate at pH 7.4. Hemoglobin concentration was adjusted to ~ 40 μM heme. B: shows the inclusion of 250 μM Fe3+ and deferiprone L1 a bidentate iron chelator.

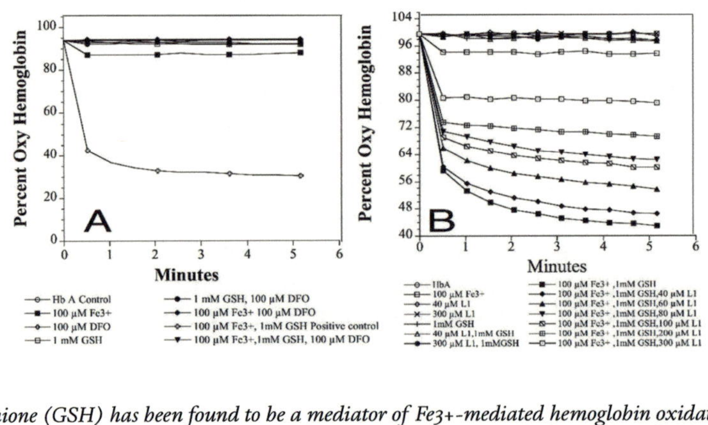

Figure 7.
Reduced glutathione (GSH) has been found to be a mediator of Fe3+-mediated hemoglobin oxidation in crude purified HbA (α2 ß2). The oxyhemoglobin content in HbA solution prepared of crude purified hemoglobin after the addition of exogenous 100 μM Fe^{3+} and 1 mM GSH is shown. In contrast to μ100 M Fe^{3+} alone, no appreciable hemoglobin oxidation occurs. A and B shows DFO and L1 at 100 μM or 1 mM reduced Fe3+ GSH-driven hemoglobin oxidation. The content of hemoglobin was adjusted to 40 M heme.

damage happens as a result of conditions, such as thalassemia and other hemoglobinopathies [6]. In these illnesses, the hemoglobin moiety's qualitative or quantitative deficiency causes iron to be ejected from the red cell into circulation [2]. Patients with thalassemia or sickle cell anemia receive blood transfusions on a regular basis as a treatment technique, which contributes to the same problem of having excess circulatory iron. This fact was examined in this study. The effect of iron on hemoglobin was investigated *in vitro* using two study models: hemolysate and pure hemoglobin solution. Winterbourn has previously demonstrated in high-throughput investigations that switching from oxyhemoglobin to methemoglobin leads to the loss of the peaks of typical normal hemoglobin spectra [7] and that is similar to what we have shown here.

From research findings, it is clear that iron has a great impact on the hemoglobin oxidation rate in hemolysate [8]. Upon adding iron in dose-dependent manner (100, 175, and 250 μM) the hemolysate constitution of oxyhemoglobin is reduced. It is also visibly seen that the hemolysate is losing the normal red pigmentation of the hemoglobin into brown color due to the iron effect of different concentrations to time (5 minutes). RBCs have been the most ferruginous cells in the human body, where the single circulating RBC contains ~ 20 mM iron, [9]. With mild hemolysis, loose iron in the body may cause the observed type of oxidative events in our body.

In contrast to hemolysate, iron does not affect purified hemoglobin. Even after 5 minutes, no change has been observed for the oxy-Hb absorbance spectra. Iron, being an oxidizing agent, requires a catalyst to produce free radicals, which are responsible for oxidative damage to hemoglobin. Free radicals are formed by enzymes, such as nicotinamide adenine dinucleotide phosphate (NADPH) oxidase, nitric oxide synthase (NOS), xanthine oxidase (XO), cytochrome P450, cyclooxygenase (COX), and lipoxygenase [10]. As evidence, **Figure 4** shows that there is no visible change in the normal red pigmentation of the hemoglobin with different iron concentrations with respect to time (5 minutes). However, when glutathione was added, there was a steep decrease in the percentage of oxyhemoglobin as a result of the pro-oxidation effect of glutathione. Similarly, it was proved from Atamna's research that reduced glutathione (GSH) can degrade heme in solution with a pH of 7 [11].

5. Conclusions

These findings might be related to the intra-erythrocytic chemical alterations seen in thalassemic individuals. When there is a quantitative deficiency in the alpha or beta chains, membrane-bound iron becomes free and can react with reduced glutathione, as we have shown. Perhaps, this may be one of the reasons, thalassemic cells further damage and hemolysis, leading to severe anemia in thalassemia [12]. However, iron chelators, such as deferoxamine (DFO) or deferiprone (L1), inhibited the GSH/Fe^{3+}-mediated hemoglobin oxidation damage.

Acknowledgements

This study was supported by "Qatar University internal grant QUUG-CAS-DHS- 15\16-21." The author acknowledges the support of students Anjud Khamis Al-Mohannadi, Israa Mahmoud Moursi, and laboratory technician Jaufar Pakari.

Author's notes

The author previously worked at Qatar University.

Author details

Ibrahim Mustafa
College of Health Science, Higher Colleges of Technology, Sharjah Women's College, Sharjah, United Arab Emirates

*Address all correspondence to: imustafa@hct.ac.ae

References

[1] Bhattacharya PT, Misra SR, Hussain M. Nutritional aspects of essential trace elements in oral health and disease: An extensive review. Scientifica. 2016;**2016**:e5464373. DOI: 10.1155/2016/5464373. Available from: https://www.ncbi.nlm.nih.gov/pmc/articles/PMC4940574/

[2] Scott MD, Eaton JW. Thalassaemic erythrocytes: Cellular suicide arising from iron and glutathione-dependent oxidation reactions? British Journal of Haematology. 1995;**91**(4):811-819

[3] Ghosh A, Banerjee S, Mitra A, Muralidharan M, Roy B, Banerjee R, et al. Interaction of p-benzoquinone with hemoglobin in smoker's blood causes alteration of structure and loss of oxygen binding capacity. Toxicology Reports. 2016;**3**:295-305

[4] Breuer W, Ermers MJ, Pootrakul P, Abramov A, Hershko C, Cabantchik ZI. Desferrioxamine-chelatable iron, a component of serum non-transferrin-bound iron, used for assessing chelation therapy. Blood. 2001;**97**(3):792-798

[5] Rossi NAA, Mustafa I, Jackson JK, Burt HM, Horte SA, Scott MD, et al. In vitro chelating, cytotoxicity, and blood compatibility of degradable poly(ethylene glycol)-based macromolecular iron chelators. Biomaterials. 2009;**30**(4):638-648

[6] Ghone RA, Kumbar KM, Suryakar AN, Katkam RV, Joshi NG. Oxidative stress and disturbance in antioxidant balance in beta thalassemia major. Indian Journal of Clinical Biochemistry. 2008;**23**(4):337-340

[7] Winterbourn CC. Oxidative reactions of hemoglobin. Methods in Enzymology. 1990;**186**:265-272

[8] Mustafa I, Hadwan TAQ. Hemoglobin oxidation in stored blood accelerates hemolysis and oxidative injury to red blood cells. Journal of Lab Physicians. 2020;**12**(4):244-249

[9] Ponka P, Sheftel AD. Erythroid iron metabolism. In: Anderson GJ, McLaren GD, editors. Iron Physiology and Pathophysiology in Humans. Humana Press; 2012. pp. 191-209. DOI: 10.1007/978-160327-485-2_10

[10] Luchi Y. Anemia Caused by Oxidative Stress. London, UK: IntechOpen; 18 Nov 2022. Available from: https://www.intechopen.com/chapters/30539

[11] Atamna H, Ginsburg H. Heme degradation in the presence of glutathione. A proposed mechanism to account for the high levels of non-heme iron found in the membranes of hemoglobinopathic red blood cells. The Journal of Biological Chemistry. 1995;**270**(42):24876-24883

[12] Scott MD, van den Berg JJ, Repka T, Rouyer-Fessard P, Hebbel RP, Beuzard Y, et al. Effect of excess alpha-hemoglobin chains on cellular and membrane oxidation in model beta-thalassemic erythrocytes. The Journal of Clinical Investigation. 1993;**91**(4):1706-1712

Chapter 4

Perspective Chapter: Low Cost Immunosuppressive Strategies in Renal Transplantation

Jacob George

Abstract

Renal transplantation is the treatment modality of choice in end stage renal disease. However, in low economic countries where government or insurance funding is not available, several patients do not opt for this treatment due to financial constraints. However, there could be options of tailored immunosuppression in both initial intensive induction immunosuppression and subsequent maintenance immunosuppression and immunomodulation thereby making this modality of treatment more cost effective. This could include selective use of induction agents, lesser frequency and dosing, use of cheaper induction agents and their combination, monitoring to decide the minimal dosage and frequency required and cost effective maintenance immunosuppressive agents with dose adjustment based on blood levels.

Keywords: cost effective immunosuppression, induction immunosuppression, maintenance immunosuppression, rituximab, renal transplantation

1. Introduction

Chronic kidney disease (CKD) is defined as abnormalities of kidney structure or function for more than three months [1]. The prevalence of CKD has been increasing worldwide with presently an estimated global prevalence in the adult population of 9.1percent [2]. Once CKD occurs, there is generally a progressive decline in renal functions to end stage renal disease (ESRD). Almost 0.8% reach ESRD by which time renal replacement therapy in the form of dialysis or renal transplantation becomes necessary to sustain life [3]. Renal transplantation is the ideal form of renal replacement therapy unless there are underlying contraindications. Yet globally only 0.01% undergo renal transplantation and this may be still lower in the lower socioeconomic countries where state funding or insurance facilities are lacking [2]. In developing countries like India, where medical insurance and reimbursement facilities are available to less than 5% of the population, renal transplantation is often out of reach to the majority [4]. The main cost of renal transplantation is for immunosuppressive medication which has to be taken life long to prevent rejection. If the financial burden due to immunosuppressive drugs could be mitigated, it could aid in making renal transplantation a viable option to economically backward patients too. This could

 IntechOpen

involve the use of less costly immunosuppressive drugs with lower dose or frequency but may need added monitoring to prevent rejection.

2. Immunosuppression in renal transplantation

Immunosuppression is necessary in organ transplantation as the transplanted organ may differ in human leucocyte antigens (HLA) and will be recognized as foreign by the recipient immune system causing it to mount a response akin to destroying a invasive microbial organism which could result in rejection of the grafted organ [5]. Immunosuppression of the recipient aims to prevent rejection. As the risk and intensity of rejection is more in the early period following transplantation, a higher degree of immunosuppression may be needed then. This is called initial immunosuppression and may need addition of induction agents which could provide more intense immunosuppression [6]. After around 6 months, the risk of rejection decreases and a lower dose of immunosuppressive drugs (maintenance immunosuppression) may suffice. A higher degree of immunosuppression theoretically reduces the chance of rejection. However, this increases the risk of opportunistic infections, metabolic side effects as well as significantly raises the cost [7–9].

The following need to be considered when deciding on possible modifications in immunosuppressive strategies with a view to reduce cost, toxicity and at the same time minimizing rejection risks.

1. Is the recipient at increased risk of rejection?

The risk of rejection can be quantified depending on the immunological risk as high or low depending on several factors like recipient age, ethnicity, degree of HLA match, presence of donor specific antibodies (DSA) and delayed graft function [10]. Those without these risk factors could be considered to have low immunologic risk and may not need intense immunosuppression.

2. Can the intensity of initial immunosuppression be reduced without posing an increased risk of graft rejection?

Patients with high immunologic risk may benefit with use of high intensity immunosuppression including use of induction agents with [11]. In the preceding decades, there was a recommendation for using induction agents for all renal transplantations [12]. However it may be possible that this may not be essential in those at a lower risk especially when combined with newer antiproliferative agents like mycophenolic acid (MPA) derivatives and tacrolimus which may confer a lower baseline acute rejection risk [13].

3. What are induction agents used in transplantation?

Induction agents can be classified based on whether they deplete lymphocytes or not. The lymphocyte depleting agents include antithymocyte globulin (ATG), humanized anti-CD52 monoclonal antibodies (mAb) alemtuzumab (Campath-1H) and the murine anti-CD3 mAb Muromonab-CD3 (OKT3) which is no longer in production. The non lymphocyte depleting agents include mAbs directed against the IL-2 receptor (IL2RA) basiliximab and daclizumab, of which the latter was subsequently withdrawn in 2009 [14].

Rituximab is a monoclonal antibody targeting B cells. Rituximab has been successfully tried as induction in highly sensitized patients awaiting renal transplantation [15], as well as in ABO-incompatible renal transplantation [16]. The efficacy of a single dose of rituximab as induction has also been reported [17].

4. Is high cost induction immunosuppression mandatory in all cases?

As acute rejection episodes are thought to have a bearing on successful outcome in renal transplantation, induction agents like IL2RA in mild immunologic risk and ATG in those with high immunologic risk were recommended as they were presumed to decrease acute rejection [11, 18]. However, most induction agents in conventional doses and frequency substantially add to the cost. The cost of two doses of IL2-RA is approximately INR 1, 00,000/−(US$ 1333) and ATG costs INR 33524 for a single dose of 50 mg (US$ 447). Conventional dosage of lymphocyte depleting agents often needs Cytomegalo virus (CMV) prophylaxis with drugs like Valganciclovir [19]. Cost of 450 mg of Valganciclovir if used for prophylaxis for 100 days is around INR 25,000-(US$ 333). Moreover the benefit of induction therapy with IL2RA has not been shown to be superior to no induction when tacrolimus was used instead of cyclosporine along with MPA thereby conferring a lower risk of acute rejection. ATG use with steroids was associated with 22% and in the setting of steroid withdrawal 27% reductions in the risk of acute rejection compared with IL-2RA, with no effect on graft survival [20]. It was observed that in standard-risk recipients on tacrolimus and MPA-based triple maintenance therapy, the addition of induction therapy with IL2RA or ATG achieved an absolute risk reduction for acute rejection of 1–4% but with no improvement in graft or patient survival. Others have also reported no benefit from use of IL-2RA induction with respect to acute rejection or graft survival [13].

5. Can induction agents be combined?

It may be possible to combine induction agents acting at different sites. ATG targets T cells and as Rituximab targets B cells, there may be a role in combining both [9, 21, 22]. Though a combination of ATG and IL2Ra has not been recommended [12], there are reports of combining ATG or IL2RA with rituximab in selected patients [9, 21, 22].

6. Is there a rationale in combining induction agents?

T cell depleting agents could decrease the risk of acute rejections [11, 18]. However, this benefit is mainly for early graft loss whereas long term graft survival has not improved [23]. Similarly the benefit of IL2RA is mainly with respect to decreasing acute cellular rejection (ACR) with even reports of increasing prevalence of antibody mediated rejection (AMR) with their usage [24]. This could be due to their less effective targeting of B cells. Rituximab targets the CD20 B cells and has been used in treating AMR [25], as well as in preventing AMR [26]. This action has been extrapolated to its use prior to ABO incompatible renal transplantation [16]. As anti donor antibodies are the main barrier to long term graft loss, it is possible that ATG alone which primarily target T cells may not have the desired effect in preventing occurrence of DSAs [27]. As Rituximab targets B cells and has been used in the management of antibody mediated rejections, there could be a role of combining it with a T cell depleting agent or IL2RA. There are reports on its use as an induction agent in combination with ATG or IL2RA albeit with differing doses and frequency [9, 21, 28].

7. Can a reduced dose and frequency of induction immunosuppression be considered?

As renal transplantation offers the best form of treatment to patients with end stage renal disease, cost cutting measures are often used in resource stretched countries with induction agents being avoided in patients with low immunologic risk. Dose of ATG as induction has traditionally been 1.5 mg/kg per day for 7 days [11]. Reducing the cumulative dose to 3–4.5 mg/kg has also been used [29]. A single dose of 50–75 mg of ATG has been used successfully in 98 Indian patients with ACR occurring in seven, combined ACR and AMR in one and hyperacute rejection in one. One patient needed a second dose of ATG due to steroid resistant ACR [4]. This suggests efficacy of even a single dose of ATG with suppression of CD3 and CD4 lymphocyte count with 1 mg/kg of ATG for upto 4–7 days. Others have also reported similar results with a lesser occurrence of AMR and lymphocyte counts being supressed in the majority by day 3 with even for 5 days in some [9]. This could suggest that daily doses of ATG for upto 7 days may not be essential if there is adequate lymphocyte suppression. Monitoring of daily peripheral lymphocyte counts, and if needed CD3 and CD4 counts and deciding need for further doses based on that may help to avoid further doses with reduction in infection episodes and cost benefit.

The dose of Rituximab in nephrology practice has often been extrapolated from the higher dose and frequency used to treat lymphoproliferative disorders [30]. A single and lower dose of 100 and 200 mg was successfully used in ABO incompatible transplants [16]. A single dose of 375 mg/m2 was effective in reducing the incidence of AMR in immunologically high-risk patients [31]. It has been shown that doses as low as 100 mg in renal transplantation while on other immunosupressants can produce prolonged CD19B cell suppression for even a year justifying avoiding the weekly doses used in lymphoproliferative disorders [22]. The cost of 100 mg Rituximab is INR3000 (US$40) making it one of the potentially cheapest induction agent. Monitoring CD 19 counts every 1–2 months subsequently and giving boosters if needed could limit the frequency of rituximab and cost.

It has also been reported that the number of CMV infections increase when the total ATG dose exceeds 7 mg/kg with the incidence of CMV being as high as 82% [32]. It is possible that valganciclovir prophylaxis may not be essential if low dose ATG is used especially if the recipient is has pre-existing CMV antibodies. Valganciclovir in a reduced dose of 450 mg on alternate days for 6 weeks has been with low dose ATG with cost benefits [4]. This could not only lessen the overall cost but more importantly decrease morbidity and mortality.

When rituximab was combined with ATG for patients with panel reactive antibody levels >50%, no rejections were seen compared to 30% cellular rejection and 26% humoral rejection when rATG was used alone [28]. This suggests a role for combining low dose rituximab with low dose ATG in those with moderate immunologic risk. In the Covid pandemic, it has been suggested that avoidance of over immunosuppression should be considered [33]. It has also been suggested that antiproliferative agents and Tcell depleting agents should be used with caution [34]. As Rituximab affects the B cells its effect on covid may not be significant. However as it can affect antibody production, it may be theoretically better to vaccinate them against Covid infection and give rituximab once protective antibodies develop. Future studies may throw light on these.

Using rituximab alone as an induction agent along with tacrolimus and mycophenolate mofetil may suffice in those with mild immunologic risk as rejection rates particularly as ACR has declined decreasing the impact of conventional induction therapy [20, 35], while AMR has increased [24] where rituximab may have a preventable role [28]. The majority of ACR in the tacrolimus era respond to 3 pulses of methyl prednisolone and only minority who do not respond need ATG [36]. As AMR responds less to therapy compared to ACR, preventing AMR may be more important suggesting a role for rituximab [37].

8. Can the dose of maintenance immunosuppressive drugs be reduced without significantly increasing the risk of rejection?

Maintenance immunosuppression protocol in most centers includes use of prednisolone, antiproliferative agent like MPA derivatives and calcineurin inhibitors (CNI) like tacrolimus or cyclosporine (CsA) [38]. The first calcineurin inhibitor used was CsA and was introduced in the early 1980sand made a significant improvement in long term graft survival [39]. However the initially suggested dose was 17 mg/kg and was associated with a risk of renal failure and other evidence of cyclosporine toxicity [40]. Subsequently dose was adjusted based on improved glomerular filtration rate (GFR) and blood pressure measurements using a target of fifty percent of the standard area-under-the-curve (AUC) dose [41]. This resulted in a significant reduction in the CsA dose. Current trials suggest better graft survival with tacrolimus compared to CsA [38]. Adjusting the CNI levels to the lower threshold recommended can be tried in patients without history of acute rejections in the first six months, stable renal functions and low immunologic risk with cost benefits, reduction of CNI toxicity including nephrotoxicity with lesser risk of infections.

Since CNIs are metabolized by the cytochrome P450 enzyme, combining it with drugs that inhibit the cytochrome P-450 system like ketoconazole, erythromycin or calcium-channel blockers could lead to higher blood levels of CNI. This strategy has been employed with significant cost benefits [42].

9. Is there a role for cost saving maintenance immunosuppression?

The main cost of maintenance immunosuppression is due to antiproliferative agents like MPA derivatives like mycophenolate mofetil (MMF) and mycophenolate sodium and CNIs. Azathioprine (Aza) was the initial antiproliferative agent [43] until MMF became commercially available in 1995 [44]. Initial reports suggesed lesser rejection episodes with MMF compared to Aza [45], with a 30–50% reduced 6-month cumulative rates of acute kidney graft rejection [46]. However, this has not been replicated in other studies where in deceased donor kidney transplant recipients on low-dose CsA and no steroids, MMF had no significant benefits over Aza [47].

The cost of Aza is approx. INR 6–18 for a day while the daily expenditure for MMF is almost ten times higher (INR 60–180; US$ 1–2.25). Using Aza could thus significantly reduce the cost as been observed by others without significantly altering rejection rates [48]. Even if some reports of lesser rejection episodes with MMF compared to Aza are considered, given that the risk of rejection episodes are less after 6 months, there could be a cost advantage if MMF is substituted with Aza after 1 year in patients with stable graft functions and without preceding acute rejections.

10. Are there laboratory parameters which could suggest reduction in dosage and frequency of immunosuppressive drugs?

It is possible to titrate the dose of several of immunosuppressive agents depending on their therapeutic blood levels, effects on cell counts and other parameters

1. Antithymocyte globulin

It has been suggested that the traditionally recommended dose of ATG could be reduced by targeting CD3 counts could prevent unnecessary higher doses [49]. Others have also reported this method as useful, reliable, and cost effective [50]. As ATG reduces the lymphocyte counts, it is possible to decide on need and timing of the subsequent dose of ATG following an initial dose based on lymphocyte counts. If they are suppressed following the initial dose, subsequent doses of ATG could be deferred till the counts normalize. Titrating further doses of ATG based on lymphocyte count alone in the absence of monitoring facilities for CD lymphocyte counts with further doses of ATG given only if absolute lymphocyte count exceeded 750/μL has also been reported [9]. This may result in a lesser number of ATG doses and a lesser cumulative dose of ATG needed [51].

2. CNIs

Titrating the area under the curve (AUC) for cyclosporine or the trough (C0) and and peak levels have been used to decide on the dose which would provide adequate immunosuppression, This could help in reducing the dose to a trough 50–100 ng/mL with the same effect as standard trough of 150–300 ng/mL [38]. Assessing the genotype of cytochrome 450 enzyme system could aid in deciding the adequate dose needed to achieve adequate blood levels with CYP3A5 expressers having a ~ 40–50% higher dose requirement compared to non-expressers. This could avoid unnecessary higher dose of CNIs like tacrolimus in some patients [52].

3. Mycophenolate mofetil

Measuring MMF levels are not as commonly used as CNI levels in renal transplantation though there are reports of adjusting the dose of MMF based on blood levels. This could result in a lesser dose of MMF without compromising on the efficacy. An AUC of 30-60 mg/hour/l at 0–12 hours is recommended to prevent renal allograft rejection [53]. This method of titrating the dose of MMF based on these levels can help to maintaining effective immunosuppression while avoiding overdosing [54].

4. Rituximab

As Rituximab is a monoclonal antibody targeting B cells, its efficacy is reflected by measuring CD19 B cells which costs approx. INR 3000 (US$ 37.5). Monitoring CD19 counts suggests a lower initial dose may suffice and targeting subsequent doses depending on B cell repopulation can result in lesser boosters and at time intervals ranging from 6 to 12 months [22]. Targeting initial dose and deciding on need for further doses depending on the CD 19 count has been found to be effective with limiting state of immunosuppression [55]. This could have significant cost advantages [22, 56].

When considering the option of lowering maintenance immunosuppression in renal transplantation, the trade-off of an increased risk of rejection should be anticipated and methods to prevent it as well as timely detection for early treatment should be ensured. Recipients with a low risk of rejection based on their immunologic risk profile, absence of previous rejection episodes and those with stable renal functions could be considered for lower immunosuppression strategies. Screening for donor specific antibodies (DSA) and protocol biopsies could provide additional information on the risk of development of future rejections and subsequent graft loss. Absence of DSA and protocol biopsies showing no evidence of active rejection could justify use of lower immunosuppression. Such patients should be advised the need for periodic check-up of renal parameters, stressed the need for drug compliance and to report promptly if any symptoms or laboratory evidence of renal dysfunction occur as response to antirejection therapies are most effective if detected early. Lowering immunosuppression has multiple advantages like lesser chances for infection, decreased risk of metabolic abnormalities and malignancies and could have advantages in the covid pandemic [9]. Multi centre trials randomizing selected patients to lower and conventional immunosuppressive protocols may shed more light on the clinical utility of lower immunosuppression strategies.

3. Conclusion

Reducing the cost of immunosuppression can help to make renal transplantation financially viable to low cost economies without state funded insurance schemes. As this could lead to increased risk of graft rejection, these strategies may need to be restricted to those committed for regular follow up and drug compliance with low immunological risk.

Acknowledgements

The assistance of residents and staff in the department of Nephrology, Government Medical College, Thiruvananthapuram in preparing this manuscript is gratefully acknowledged.

Funding Statement

This study has not received any funding from any source.

Conflict of interest statement

The author reports no conflict of interest in this manuscript.

Author details

Jacob George
Nephrology, Government Medical College, Thiruvananthapuram, India

*Address all correspondence to: drjacobgeo@gmail.com

References

[1] Levey AS, Eckardt KU, Dorman NM, et al. Nomenclature for kidney function and disease: Report of a kidney disease: Improving global outcomes (KDIGO) consensus conference. Kidney International. 2020;**97**:1117-1129

[2] Chronic Kidney Disease Collaboration. Global, regional, and national burden of chronic kidney disease, 1990-2017: A systematic analysis for the global burden of disease study 2017. Lancet. 2020;**395**:709-733

[3] Singh AK, Farag YM, Mittal BV, et al. Epidemiology and risk factors of chronic kidney disease in India - results from the SEEK (screening and early evaluation of kidney disease) study. BMC Nephrology. 2013;**14**:114

[4] Abraham G, Vijayan M, Parthasarathy R, et al. Single dose thymoglobulin induction therapy in prevention of acute rejection in renal transplant recipients. Journal of Nephropharmacology. 2018;**7**(1):32-35

[5] Adler SH, Turka LA. Immunotherapy as a means to induce transplantation tolerance. Current Opinion in Immunology. 2002;**14**(5):660-665

[6] Hill P, Cross NB, Barnett ANR, Palmer SC, Webster AC. Polyclonal and monoclonal antibodies for induction therapy in kidney transplant recipients. Cochrane Database of Systematic Reviews. 11 Jan 2017;**1**(1):1-10

[7] Fishman JA. Infection in solid-organ transplant recipients. The New England Journal of Medicine. 2007;**357**(25):2601-2614

[8] van Sandwijk MS, Bemelman FJ, Ten Berge IJ. Immunosuppressive drugs after solid organ transplantation. The Netherlands Journal of Medicine. 2013;**71**(6):281-291

[9] George J, Gracious N, Gopal A, et al. Low-dose induction immunosuppression in deceased donor kidney transplantation during coronavirus disease pandemic - a multicentric prospective observational study. Indian Journal of Transplant. 2021;**15**:313-319

[10] Pratschke J, Dragun D, Hauser IA, et al. Immunological risk assessment: The key to individualized immunosuppression after kidney transplantation. Transplantation Reviews. 2016;**30**(2):77-84

[11] Brennan DC, Daller JA, Lake KD, et al. Rabbit antithymocyte globulin versus basiliximab in renal transplantation. The New England Journal of Medicine. 2006;**355**:1967-1977

[12] Kidney Disease: Improving Global Outcomes (KDIGO) Transplant Work Group. KDIGO clinical practice guideline for the care of kidney transplant recipients. American Journal of Transplantation. 2009;**9**(Suppl 3):S1-S155

[13] Hellemans R, Bosmans JL, Abramowicz D. Induction therapy for kidney transplant recipients: Do we still need anti-IL2 receptor monoclonal antibodies? American Journal of Transplantation. 2017;**17**(1):22-27

[14] Wiseman AC. Induction therapy in renal transplantation: Why? What agent? What dose? We may never know. Clinical Journal of the American Society of Nephrology. 2015;**10**(6):923-925

[15] Sood P, Hariharan S. Anti-CD20 blocker rituximab in kidney

transplantation. Transplantation. 2018;**102**(1):44-58

[16] Hatakeyama S, Fujita T, Murakami R, et al. Outcome comparison of ABO-incompatible kidney transplantation with low-dose rituximab and ABO-compatible kidney transplantation: A single-center experience. Transplantation Proceedings. 2014;**46**:445-448

[17] Tydén G, Genberg H, Tollemar J, Ekberg H, Persson NH, Tufveson G, et al. A randomized, Doubleblind, placebo-controlled, study of single-dose rituximab as induction in renal transplantation. Transplantation. 2009;**87**(9):1325-1329

[18] Cosimi A. The clinical value of antilymphocyte antibodies. Transplantation Proceedings. 1981;**13**:462-468

[19] Mourad G, Garrigue V, Squifflet JP, et al. Induction versus noninduction in renal transplant recipients with tacrolimus-based immunosuppression. Transplantation. 2001;**72**(6):1050-1055

[20] Tanriover B, Zhang S, MacConmara M, et al. Induction therapies in live donor kidney transplantation on tacrolimus and Mycophenolate with or without steroid maintenance. Clinical Journal of the American Society of Nephrology. 2015;**10**:1041-1049

[21] Pathak V, Madhavan D, Narayanasamy K, et al. Low-dose rituximab and thymoglobulin induction with steroid-free maintenance immunosuppression and protocol biopsies improves long-term patient and graft survival after kidney transplantation: Survival and safety outcomes in more than 1100 patients from a single center. Transplantation direct. 2019;**5**:e475

[22] George J, Alex S, Thomas ETA, et al. Clinical response and pattern of B cell suppression with single low dose rituximab in nephrology. Kid. 2020;**360**(1):359-367

[23] Nankivell BJ, Kuypers DR. Diagnosis and prevention of chronic kidney allograft loss. The Lancet. 2011;**378**:1428-1437

[24] Bamoulid J, Staeck O, Crepin T, et al. Anti-thymocyte globulins in kidney transplantation: Focus on current indications and long-term immunological side effects. Nephrology, Dialysis, Transplantation. 2017;**32**:1601-1608

[25] Roberts DM, Jiang SH, Chadban SJ. The treatment of acute antibody-mediated rejection in kidney transplant recipients-a systematic review. Transplantation. 2012;**94**:775-783

[26] Macklin PS, Morris PJ, Knight SR. A systematic review of the use of rituximab for the treatment of antibody-mediated renal transplant rejection. Transplantation Reviews. 2017;**31**(2):87-95

[27] Colvin RB, Smith RN. Antibody-mediated organ-allograft rejection. Nature Reviews Immunology. 2005;**10**:807-817

[28] Laftavi MR, Pankewycz O, Feng L, et al. Combined induction therapy with rabbit antithymocyte globulin and rituximab in highly sensitized renal recipients. Immunological Investigations. 2015;**44**(4):373-384

[29] Singh N, Rossi AP, Savic M, et al. Tailored rabbit antithymocyte globulin induction dosing for kidney transplantation. Transplantation direct. 2018;**4**:e343-e349

[30] McLaughlin P, Grillo-Lopez AJ, Link BK, et al. Rituximab chimeric

anti-CD20 monoclonal antibody therapy for relapsed indolent lymphoma: Half of patients respond to a four dose treatment program. Journal of Clinical Oncology. 1998;**16**:2825-2833

[31] van den Hoogen MWF, Steenbergen EJ, Baas MC, Florquin S, Hilbrands LB. Absence of Intragraft B cells in rejection biopsies after rituximab induction therapy: Consequences for clinical outcome. Transplantation Direct. 2017;**3**(4):e143

[32] Clesca P, Dirlando M, Park SI, et al. Thymoglobulin and rate of infectious complications after transplantation. Transplantation Proceedings. 2007;**39**(2):463-464

[33] Martino F, Plebani M, Ronco C. Kidney transplant programmes during the COVID-19 pandemic. The Lancet Respiratory Medicine. 2020;**8**:e39

[34] Maggiore U, Abramowicz D, Crespo M, et al. How should I manage immunosuppression in a kidney transplant patient with COVID-19? An ERA-EDTA DESCARTES expert opinion. Nephrology, Dialysis, Transplantation. 2020;**35**:899-904

[35] Kamel M, Kadian M, Srinivas T, et al. Tacrolimus confers lower acute rejection rates and better renal allograft survival compared to cyclosporine. World Journal of Transplantation. 2016;**6**(4):697702

[36] Mazzucchi E, Lucon AM, Nahas WC, et al. Histological outcome of acute cellular rejection in kidney transplantation after treatment with methylprednisolone. Transplantation. 1999;**67**(3):430-434

[37] Lefaucher C, Nochy D, Hill GS, et al. Determinants of poor graft outcome with antibody mediated acute rejection. American Journal of Transplantation. 2007;**7**:832-841

[38] Ekberg H, Tedesco-Silva H, Demirbas A, et al. ELITE-symphony study. Reduced exposure to calcineurin inhibitors in renal transplantation. The New England Journal of Medicine. 2007;**357**(25):2562-2575

[39] Harder F, Loertscher R, Thiel G. Cyclosporin in cadaveric renal transplantation: One-year follow-up of a multicentre trial. The Lancet. 1983;**8357**(2):986-989

[40] Calne RY, White DJG, Evans DB. Cyclosporin a in cadaveric organ transplantation. British Medical Journal. 1981;**282**:934-936

[41] Etienne O, Toupance J, B′enichou et al. "A 50% reduction in cyclosporine exposure in stable renal transplant recipients: Renal function benefits." Nephrology Dialysis Transplantation. 2010;**25**: 3096-3106

[42] Kahan BD. Drug therapy: Cyclosporine. The New England Journal of Medicine. 1989;**321**:1725-1738

[43] Zukoski CF, Lee HM, Hume DM. The prolongation of functional survival of canine renal homografts by 6-mercaptopurine. Surgical Forum. 1960;**11**:470-472

[44] Sollinger HW. Mycophenolate mofetil for the prevention of acute rejection in primary cadaveric renal allograft recipients. U.S. renal transplant Mycophenolate Mofetil study group. Transplantation. 1995;**60**(3):225-232

[45] Halloran P, Mathew T, Tomlanovich S, Groth C, Hooftman L, Barker C. Mycophenolate mofetil in renal allograft recipients: A pooled efficacy analysis of three randomized,

double-blind, clinical studies in prevention of rejection. The international Mycophenolate Mofetil renal transplant study groups. Transplantation. 1997;**63**(1):39-47

[46] The Tricontinental Mycophenolate Mofetil Renal Transplantation Study Group. A blinded, randomized clinical trial of mycophenolate mofetil for the prevention of acute rejection in cadaveric renal transplantation. Transplantation. 1996;**61**:1029-1037

[47] Ruggenenti P, Cravedi P, Gotti E, et al. Mycophenolate mofetil versus azathioprine in kidney transplant recipients on steroid-free, low-dose cyclosporine immunosuppression (ATHENA): A pragmatic randomized trial. PLoS Medicine. 2021;**18**(6):e100366

[48] Remuzzi G, Cravedi P, Costantini M, et al. Mycophenolate mofetil versus azathioprine for prevention of chronic allograft dysfunction in renal transplantation: The MYSS follow-up randomized, controlled clinical trial. Journal of the American Society of Nephrology. 2007;**18**:1973-1985

[49] Yilmaz M, Sezer TO, Kir O, Ozturk A, Hoscoskun C, Toz H. Use of ATG-Fresenius as an induction agent in deceased-donor kidney transplantation. Transplantation Proceedings. 2017;**49**:486-489

[50] Ata P, Kara M, Özdemir E, Canbakan M, Gökçe AM, Bayraktar FA, et al. Monitoring of CD3(+) T-cell count in patients receiving antithymocyte globulin induction after cadaveric renal transplantation. Transplantation Proceedings. 2013;**45**(3):929-931

[51] Klem P, Cooper JE, Weiss AS, et al. Reduced dose rabbit anti-thymocyte globulin induction for prevention of acute rejection in high-risk kidney transplant recipients. Transplantation. 2009;**88**:891-896

[52] Niioka T, Kagaya H, Saito M, et al. Impact of the CYP3A5 genotype on the distributions of dose-adjusted trough concentrations and incidence of rejection in Japanese renal transplant recipients receiving different tacrolimus formulations. Clinical and Experimental Nephrology. 2017;**21**(5):787-796

[53] van Hest RM, Mathot RA, Vulto AG, et al. Within-patient variability of mycophenolic acid exposure: Therapeutic drug monitoring from a clinical point of view. Therapeutic Drug Monitoring. 2006;**28**(1):31-34

[54] Joy MS, Hilliard T, Hu Y, et al. Influence of clinical and demographic variables on mycophenolic acid pharmacokinetics in antineutrophil cytoplasmic antibody-associated vasculitis. The Annals of Pharmacotherapy. 2009;**43**:1020-1027

[55] Cravedi P, Ruggenenti P, Sghirlanzoni MC, Remuzzi G. Titrating rituximab to circulating B cells to optimize lymphocytolytic therapy in idiopathic membranous nephropathy. Clinical Journal of the American Society of Nephrology. 2007;**2**:932-937

[56] Ruggenenti P, Chiurchiu C, Abbate M, Perna A, Cravedi P, Bontempelli M, et al. Rituximab for idiopathic membranous nephropathy: Who can benefit? Clinical Journal of the American Society of Nephrology. 2006;**1**:738-748

Chapter 5

Perspective Chapter: Immunosuppression in Patients with Diabetes Mellitus

Pratima Tripathi

Abstract

Diabetes is an age-dependent health issue prevalent worldwide and specially seen in those families with prevalent history of the disorder. Insufficient insulin production by the defective pancreas that leads to high blood glucose levels in the systemic circulation makes the patients more prone to an infection that exaggerates with time as compared to their counterparts. This increased prevalence of infections in diabetics may be due to defects in the immune functionality of the diabetes patients. High blood glucose level evokes inflammatory responses due to provoked inflammatory immune response against hyperglycemic condition in adipocytes and macrophages. The inflammatory mediators attack the pancreatic beta cells thus affecting the insulin production, which in-turn again results in hyperglycemia. Dysfunction of the immune response could not control the invasion of pathogens thereby, increasing the incidence of infectious diseases and related co- morbidities. This chapter discusses about immune dysfunction and suppression in T2DM and the underlying inflammation and infections in diabetics. An elaborate and in-depth understanding of the immune dysfunction in T2DM patients can help in the management and development of better targeted therapeutics to cure the disorder. It may also provide an insight in how to take care of one's health as a precautionary measure to avoid the complications leading to diabetes and vice versa.

Keywords: type 2 diabetes, hyperglycemia, immune dysfunction, immune suppression, inflammation and infection

1. Introduction

Diabetes caused by chronic hyperglycemia due to failure of the pancreatic beta cells to produce adequate insulin or ineffective utilisation of the produced insulin by the body is a severe health issue worldwide [1]. Diabetes exists in two major forms: type 1 (T1D) and type 2 (T2D) diabetes. Type 1 diabetes is caused by the body's immune system damaging the pancreas thereby making the body incapable to produce sufficient insulin. Impaired regulation and use of glucose due to insulin resistance or inefficiency in insulin production by the body leads to Type 2 diabetes. Family history is the well risk factors for Type 1 diabetes whileobesity, advancing age, family history, sedentary lifestyle, ethnicity and certain medications are the risk factors associated with Type 2 diabetes. Diabetes affects the brain, kidney, heart, and eyes as an acute condition that rises the risk of various diseases brought on by damage to the macro and microvasculature [2]. Patients with diabetes are also more prone to infections. Numerous studies have shown that individuals with diabetes are more likely to develop diseases of the lower respiratory tract, including pulmonary tuberculosis (TB) urinary tract infections, and pneumonia and infections of the skin and internal organ tissues [3]. Diabetes patients typically have poor outcomes from infection treatment. Diabetes patients are more financially burdened by infection because of the high cost of therapy, the time of treatment, and the associated consequences.

Around 425 million people worldwide have diabetes, according to the International Diabetes Federation [2]. Both developed and developing nations expect this number to rise. By 2045, there could be 629 million diabetes patients worldwide if adequate care and control are not implemented. Around 5 million persons perished from diabetes in 2017; 850 million USD were spent on diabetic care. In developing nations, especially those with tropical climates have high prevalence of communicable disease, along with whooping number of diabetics, which inevitably results in increased incidence of infectious diseases posing huge burden on the nation's economy [4].

Due to decreased insulin synthesis by islet cells in the pancreas and insufficient insulin action (insulin resistance), T2D accounts for over 90% of all cases of diabetes. The condition causes the blood glucose levels to rise. Obesity, inactivity, and ageing are all linked to insulin resistance in T2D. In order to counteract insulin resistance, the pancreatic islets expand their cell mass and produce more insulin [5]. When this attempt falls short of making up for insulin resistance, T2D develops. Pancreatic cell damage due to years of insulin resistance leads above half of T2D patients to take insulin therapy. In T2D, long-term chronic insulin resistance has a number of negative effects, such as atherosclerosis and microvascular problems such nephropathy, neuropathy, and retinopathy [6].

2. Glucose intolerance and insulin resistance

Following a meal, blood islet cells produce and release insulin in response to elevated blood glucose levels. Lower blood glucose level results due to increased glucose uptake by cells as a result of the insulin binding to its receptors present on the cell membranes. This process causes the translocation of glucose transporters to the cell membrane. Hyperglycemia is a condition where the pancreas either fails to generate enough insulin, produces insufficient insulin, or both. It has been observed that TNF levels elevated in adipose tissue of obese mice have been linked to insulin resistance in these experimental models [7]. Additionally, increased levels of interleukin (IL)-6, plasminogen activator inhibitor, C-reactive protein and other inflammatory mediators in the plasma of obese mice exaggerates the damage associated with these factors. Suppression of insulin receptor substrate (IRS-1) is brought on by TNF, ceramide, diacylglyceride, free fatty acids, hypoxia, reactive oxygen species (ROS), and c-Jun N-terminal kinase I (JNK1) in liver and adipose tissue (**Figure 1**). Additionally, TNF- causes insulin resistance by impairing the activity of the gamma subunit of the peroxisome proliferator-activated receptor [8].

Tyrosine phosphorylation at IRS-1 and -2 results from the binding of insulin with its receptor. IKK and JNK1, the mediators of inflammatory and stress responses, phosphorylate IRS substrates on serine, which inhibits insulin signalling. The

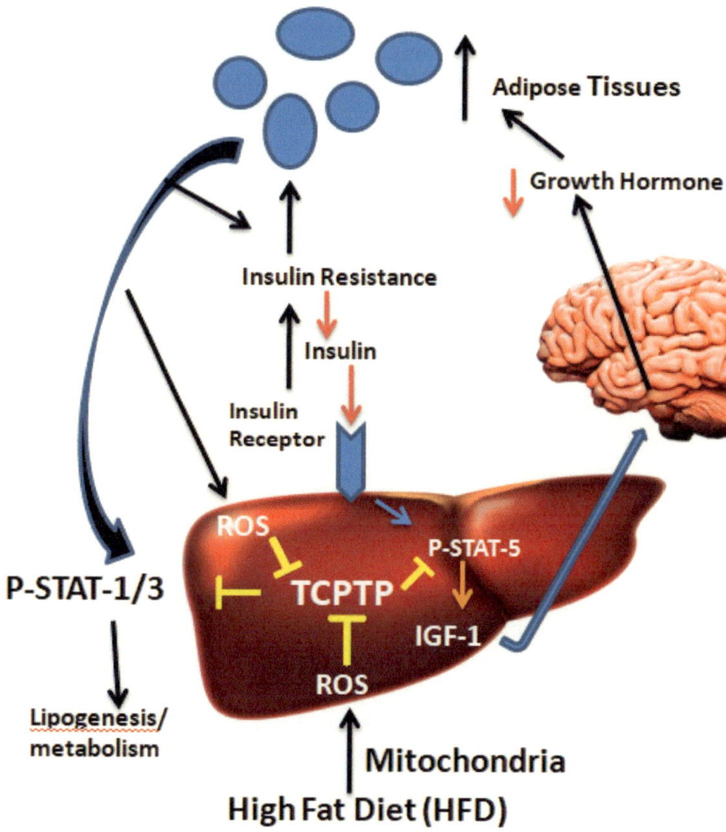

Figure 1.
Oxidative stress promotes a cascade of responses leading to adipogenesis.

transcriptional activation of several genes linked to the inflammatory response is also caused by JNK1 and IKK, which leads to insulin resistance. JNK1 and IKK signalling pathways are activated by the influx of more free fatty acids and glucose [9]. The transcription inflammation associated genes causes the phosphorylation and activation of IKK that further encourages the ubiquitination and destruction pathways in proteasome thereby translocating NF into the nucleus. IKK also blocks insulin signalling pathways in adipocytes by phosphorylating IRS-1 serine residues [10]. TNF-induced JNK activation phosphorylates IRS-1 to suppress insulin signalling. The transducers and activators of Janus kinase/signal transcription (JAK/STAT) pathway also results in the suppression of insulin signalling. STAT's tyrosine is phosphorylated by JAK kinases, which causes STAT to dimerize and go to the nucleus and phosphorylate IRS-1 at Ser636 and Ser307.The Glut-4 translocation to cell membranes is eventually hampered by this suppression of insulin signalling, which results in hyperglycemia [11].

3. Hypoinsulinemia and apoptosis of pancreatic B-cell

Crosstalk between pathogenic CD4+ and CD8+ T cells and CD11c+ M1 macrophages in obese adipose tissue further intensifies the inflammatory immune response brought on by adipocyte apoptosis and macrophage infiltration, which worsens adipose tissue inflammation and peripheral insulin resistance [12]. As a result, pancreas cells produce more insulin to offset peripheral insulin resistance, which leads to hyperinsulinemia. The causes of T2DM are multifaceted and include insulin resistance brought on by obesity, poor insulin production, and loss of cell mass due to cell death. Absolute cell insufficiency in T1DM and relative cell deficiency in T2DM are both caused by apoptosis. The TNF receptor superfamily includes Fas (CD 95), which is distinguished by having a death domain motif in the cytoplasmic terminus [13]. A membrane-bound protein called Fas L (CD 178) is increased on activated T cells. Apoptosis is considerably inhibited by the expression of dominant-negative Fas or neutralising antibodies to Fas, which also results in adequate cell function, prevents the adoptive transmission of diabetes by primed T-cells, and slows the progression of T1DM development [14].

Insulin resistance, impaired insulin production, loss of cell mass with increased cell death, and islet amyloid deposits are the hallmarks of T2DM. In T2DM, obesity-related insulin resistance is followed by a failure of beta-cell insulin production to counteract the deteriorating insulin sensitivity. A balance between beta-cell replication and apoptosis, as well as islet hyperplasia and the creation of additional islets from exocrine pancreatic ducts, regulates beta-cell mass [15]. In cells from T2DM patients, elevated caspase-3 and -8 activate, which can be reduced by the anti-diabetic drugs. The delicate balancing act between cell replication and apoptosis, which is regulated by a balance between matrix metalloproteinase (MMP)-1 and -2 and tissue inhibitor of MMP (TIMP)-1 and -2, is essential for islet development and function in vivo. The -cells undergo continual remodelling [16]. However, chronic growing insulin resistance over time finally results in exhausted beta cells and an insulin shortfall. Additionally, the build-up of free fatty acids, amyloids, and inflammatory cytokines triggers the death of beta cells, resulting in long-term hyperglycemia and T2D.

4. Propensity of infection in hyperglycemia

The immune system employs incredible defences to keep out the invading viruses, bacteria, fungi, poisons, and parasites. In healthy condition make it tough for viruses to get past this protection, but a number of illnesses and flaws make the immune system malfunction. Pus, for instance, indicates that there is an infection since bacteria can readily enter an open wound and convert it into a non-healing wound [17]. Natural barriers, such as healthy surfaces of the skin and mucosa, as well as the production of cytokines, chemokines, and ROS, aid our defence mechanisms in stopping pathogenic infiltration.

Unfortunately, diabetes disrupts the immunological response of the host. Neuropathy increases the probability of natural barrier deterioration, and T2D can also have an impact on cellular immunity. Insufficient insulin and high blood sugar are the causes of this [18]. As a result of the immune system's inability to defend against invasive microorganisms, infections are a significant problem for people with diabetes, according to the American Diabetes Association [19]. Numerous investigations have been made to identify the pathways connected to diabetes that weaken the host's defence against infections. These processes include inhibition of cytokine production, flaws in phagocytosis, immunodeficiency, and failure to eradicate microorganisms. There is a widespread perception that people with diabetes are more susceptible to infectious diseases, still very few studies have thoroughly examined the population's overall risk for infections. The first such study, conducted in Canada, looked back on the incidence of infection and/or death among diabetic patients and age-matched controls, accumulating more than 500,000 cases per group over two separate time periods. According to this data, diabetic patients have a noticeably increased prevalence of infections; the most common illnesses were bacterial infections including osteomyelitis, pyelonephritis, cystitis, pneumonia, cellulitis, sepsis, or peritonitis [20]. Another study, from Netherland, prospectively compared patients with type I or type II diabetes to patients with hypertension in terms of the frequency of certain infections. Diabetes increases the risk of bacterial skin and mucous membrane infections, urinary tract infections, and lower respiratory tract infections [21]. This is consistent with the widespread observation that diabetes individuals have an elevated risk for wound infections, most likely due to the higher prevalence of leg ulcers in these patients.

There are several other factors that connect diabetes to infections, in addition to the fact that it appears to be a separate risk factor for bacterial infections: (1) Diabetic individuals are more likely to contract specific (rare) illnesses, and (2) diabetic people are more likely to have particular consequences when exposed to pathogens. As a result, some uncommon illnesses, such as emphysematous pyelonephritis, invasive otitis externa, emphysematous cholecystitis, or rhinocerebralmucormycosis, are more common in patients with diabetes [22]. Diabetes also appears to raise the risk of infections brought on by specific bacteria, including *Staphylococcus aureus* and *Mycobacterium tuberculosis*. It was postulated to explain that increased fatality rates from pneumococcal pneumonia in these individuals are linked to infections by particular species, such as Streptococcus pneumonia thereby increasing the chances of bacteremia [18]. Additionally, a report from the Community-Acquired Pneumonia Organisation international cohort study revealed that diabetes was not a risk factor for death when suffering from bacteremic pneumococcal infection and that pneumococcal bacteremia did not affect the outcome in terms of clinical stability in patients with diabetes mellitus.

These findings fuel the debate over whether cardiovascular and renal comorbidities, which are frequently associated with diabetes, may actually increase susceptibilities to infections and affect the outcome from infections rather than the metabolic changes seen in diabetic subjects. This poses an uncertainty whether diabetes mellitus is actually a significant risk factor for important infections such as pneumonia (**Figure 2**).

4.1 Impaired cytokines in diabetes

Under hyperglycemic condition, peripheral blood mononuclear cells (PBMCs) and isolated monocytes from persons with T1D and T2D releases less interleukin 1 beta (IL-1) after being activated with lipopolysaccharides (LPS). T1D participants' monocytes derived from PBMCs produced fewer IL-1 and IL-6 as compared to healthy donors [23]. The PBMCs of non-diabetic subjects were activated by anti-CD3 antibodies, and when they were exposed to high glucose levels, they were found to reduce the production of the cytokines IL-2, IL-6, and IL-10.Because IL-6 is essential for pathogen defence as well as for adaptive immune response by inducing antibody production and effector T-cell development, studies have shown that inhibiting those cytokines in hyperglycemia may suppress the immune response against pathogens that are invading the body. Dextrose octreotide-induced PBMCs from healthy patients were shown to produce less IL-6 and IL-17A, particularly in CD14+ and CD16+ intermediate monocytes, suggesting that high blood glucose levels had an adverse effect on the functioning of the immune system [24].

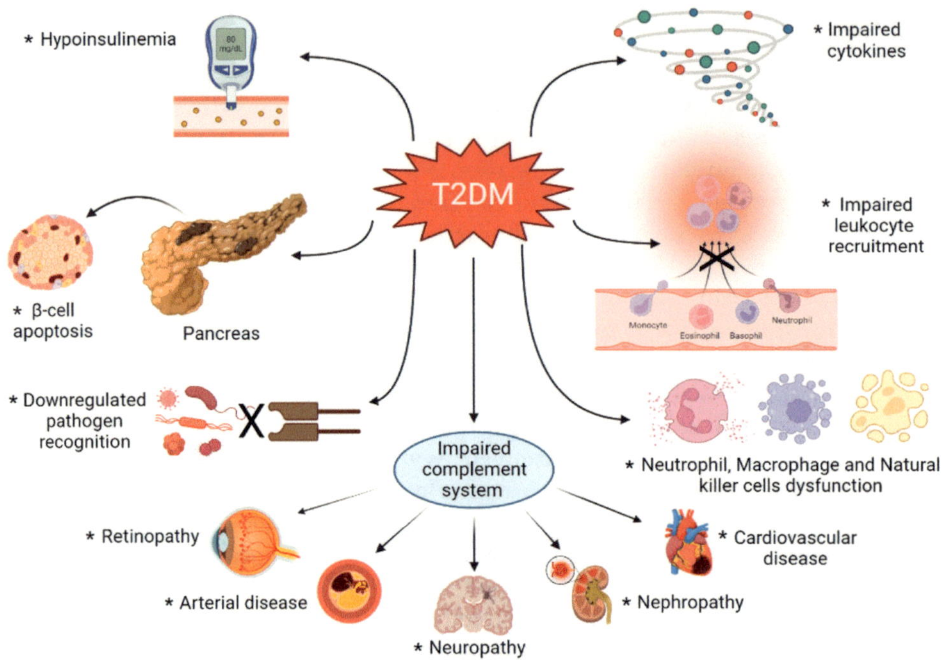

Figure 2.
Impact of T2DM on immune system leading to several associated complications.

Loss of IL-10 release by Myeloid cells' and production of interferon gamma (IFN-) and TNF by T cells' is caused by increased glycation while is decreased during diabetes. When compared to normal mice, IL-22 cytokine levels were found to be lower in obese leptin-receptor-deficient (db/db) mice and high-fat diet-induced hyperglycemic animals. In PBMC cultivated in a high glucose medium and stimulated by poly I:C, type 1 IFN production reduces [22]. Following infection with *Burkholderia pseudomallei*, an IA investigation found that in diabetes patients PBMC cultures produce less IL-12 and IFN than PBMCs from healthy donors. Additionally, PBMCs from diabetics had a greater intracellular bacterial load than those from healthy controls, indicating that hyperglycemia weakens the host's defence against bacterial invasion. Recombinant IL-12 and IFN dramatically decreases bacterial load in PBMCs from diabetic people, demonstrating that low IL-12 and IFN production in diabetes reduces the ability of immune cells to control bacterial development during infection [25]. Therefore, it is believed that diabetic hyperglycemia reduces the ability of macrophages and other leukocytes to destroy infections. The impact of insulin shortage on macrophage activity against pathogens in T2D has not been as extensively studied as the influence of hyperglycemia on immune cell activity in T2D. The infusion of insulin into bone marrow-derived macrophages isolated from diabetic mice dramatically boosts the production of TNF and IL-6 after LPS stimulation, according to research on the effects of insulin deficit on immune response [7]. Another rat investigation found that insulin deficiency disrupts the alveolar macrophage phagocytosis and cytokine production, both of which gets recovered after insulin administration. This data suggest that the injection of exogenous insulin in diabetics may boost the immune function.

4.2 Impediment of leukocyte recruitment

A robust T-cell-mediated response is essential for host defence against intracellular bacterial infections. A variety of 120 T cell subtypes (Th1, Th2, Th17, and Treg) develop into diverse immune responses that are primarily based on released cytokine patterns. A key factor in the ability to resist intracellular bacterial infections is the early influx of IFN-producing Th1 cells [26]. There is compelling evidence to suggest that diabetic hosts experience an initial delay in the activation of Th1 cell-mediated immunity. Although it could be too late to prevent diabetic hosts from bacterial spread, there is also clinical and experimental evidence suggesting the late inflammatory response during chronic TB is strengthened. It's likely that the enhanced antigenic stimulus that caused this late hyper-inflammatory response, as a result of defective innate immune regulation, or as a result of cumulative build up contributing to the chronic inflammation underlying the immunopathology of diabetes itself [27]. Patients with co-morbid diabetes and tuberculosis have been reported to have elevated levels of circulating Th1- and Th17-associated cytokines.

Leukocyte recruitment, which typically occurs in three stages, is a well-organised cascade-like process that involves (a) selectin-dependent leukocyte rolling on the endothelium layer, (b) chemokine-dependent integrin activation with subsequent leukocyte adherence, and (c) diapedesis. Much has been learnt about the transmigration process which involves the final stage of leukocyte recruitment into inflamed tissues. Leukocyte transmigration is influenced by a number of adhesion molecules, including platelet cells adhesion molecule, junctional adhesion molecule-1, and CD99, while leukocyte motility in tissues is influenced by 1-integrins [28]. The infiltration of CD8 + T cells and CD45+ leukocytes was drastically decreased in the brains of db/db mice that had West Nile virus-associated encephalitis. This study demonstrated that reduced recruitment of CD45+ leukocytes and CD8+ T lymphocytes was related to lower expression of cell adhesion molecules (CAMs), such as E-selectin and intracellular adhesion molecule (ICAM)-1 [29]. *In vivo* investigation employing streptozotocin-induced diabetic mice infected with *Klebsiella pneumoniae* likewise proved this impairment in leukocyte recruitment. Granulocyte counts in the alveolar airspace of the diabetic mice were lower. They also discovered that after inhaling *Klebsiella pneumoniae* LPS, lung tissue produced less of several cytokines, including CXCL1, CXCL2, IL-1, and TNF [30].

4.3 Erratum in pathogen recognition

Pathogen recognition receptors that play a crucial role in the innate immune system include Toll-like receptors (TLRs) and NOD-like receptors (NLRs). Different adaptor proteins, which are frequently identified to activate the NF-kB and hence stimulate the release of proinflammatory cytokines, mediate both the TLRs and NLRs pathways. The pathophysiology of inflammation-mediated insulin resistance, which further develops metabolic problems, has been hypothesised to include TLRs and NLRs significantly. Innate immunity is activated by TLR2 homodimers and TLR2 heterodimers with TLR1 or TLR6 upon detection of damage-associated molecular patterns (DAMPs) that are endogenous chemicals created and released during T2DM, an infection or inflammatory response [31]. Inflammation has been shown to play a significant role in type 2 diabetes-related pancreatic beta cell dysfunction [32]. Therefore, the inflammatory effects of the TLR2-ligand interaction may play a significant role in the development of type 2 diabetes. According to a study the interaction between TLR2/6 and its associated ligands causes macrophage activation and the generation of pro-inflammatory cytokines IL-1 and IL-6, which contribute to islet inflammation.

The expression of TLR-2 and TIRAP, which are involved in the identification of pathogens, was found to be downregulated in diabetic mice [31]. However, numerous investigations have demonstrated enhanced TLR expression in neutrophils and monocytes isolated from diabetic individuals. TLR was found to be under expressed in diabetics with poor glycemic control, but higher in patients with controlled hyperglycemia without complications [33]. Therefore, it is yet unknown how hyperglycemia affects TLR expression and associated immunity in diabetic people.

5. Debilitated neutrophils

The quick release of ROS and the presence of pre-formed proteolytic granules make neutrophils one of the first phagocytic cells to reach infection sites. Clinically, the majority of infected cells in patients with active tuberculosis' sputum and bronchoalveolar lavage are neutrophils. The results of in vitro research on neutrophils' capacity to kill *M. tuberculosis* and *B. pseudomallei* vary, which is likely due to a variety of host- and organism-specific variables as well as variations in experimental methodology [29]. Unstimulated neutrophils in diabetics have been shown to produce more inflammatory cytokines and ROS, which has been linked to AGE-direct activation. However, in diabetic hosts, neutrophil responses to infection seem to be primarily inhibited. Reduced glucose metabolism via the pentose-phosphate route, which generates NADPH, a need for adequate NADPH oxidase function, may be linked to decreased pathogen-stimulated ROS generation [28]. Furthermore, neutrophil dysfunction in diabetic hosts may be caused by diminished glutathione reductase activity, which also controls neutrophil-based ROS generation and phagocytosis. ROS induces the release of neutrophil extracellular traps (NET), another significant bactericidal mechanism in addition to directly killing germs. Such deficiencies in neutrophil function might make it easier for internal bacteria to use neutrophils as a haven and a vehicle for spreading in diabetic hosts [34].

Following stimulation with phorbol 12-myristate-13-acetate, isolated neutrophils from T2D TB patients produced less ROS. Increased resistin levels in the blood of T2D patients were linked to this ROS generation deficiency [35]. When exposed to a high glucose content media, isolated neutrophils from healthy patients suppressed superoxide (O2−). Through the suppression of glucose-6-phosphate dehydrogenase (G6PD), which interfered with nicotinamide adenine dinucleotide phosphate synthesis, this impairment was caused.

When healthy individuals' blood was exposed to bacterial wall components after becoming hyperglycemic the blood's neutrophil degranulation reduces. Another example of neutrophil dysfunction in *S. aureus* phages was provided by C3-mediated complement suppression brought on by hyperglycemia [36]. It is reported that NETs, which increase susceptibility to infections, are less likely to form when hyperglycemia is present. All of these studies showed that hyperglycemia causes neutrophil dysfunction, which includes irregularities in ROS production, impairments in neutrophil degranulation, inhibition of immunoglobulin-mediated opsonization decreased phagocytosis, and errors in NET formation.

6. Dysfunctional macrophages

Early host defence against intracellular bacterial infections is greatly aided by macrophages. In order to control infection, macrophages perform crucial effector tasks such as phagocytosing pathogens and eliminating necrotic and apoptotic neutrophils. Activation and recruitment of circulating monocytes to infection sites, where they undergo macrophage differentiation, are aided by cytokines and chemokines produced by neutrophils, such as TNF- and CCL2 [37]. The macrophage cytokine profile is essential for promoting efficient cell-mediated immunity and defence against intracellular bacteria in addition to phagocytic and antibacterial processes. Inducible nitric oxide synthase, co-stimulatory molecules, and inflammatory cytokines like TNF-, IL-12, and IL-18 are all up-regulated as a result of M1 macrophage polarisation in response to intracellular bacterial infections. IFN- production from NK cells depends on the production of IL-12 and IL-18 [38]. Inducible nitric oxide synthase, co-stimulatory molecules, and inflammatory cytokines like TNF-, IL-12, and IL-18 are all up-regulated as a result of M1 macrophage polarisation in response to intracellular bacterial infections. In order to create T helper type 1 (Th1) cell-mediated immunity, NK cells and T cells must produce IL-12 and IL-18 in order to trigger an IFN- response [37]. Both IFN and TNF stimulate inducible nitric oxide synthase and NADPH oxidase, which activate macrophages and aid in the destruction of intracellular microorganisms.

The way that macrophages work is also changed by hyperglycemia. Chronic hyperglycemia gets significantly correlated with deficiencies in complement receptors and Fc receptors on isolated monocytes, impairing phagocytosis [39]. Reduced phagocytosis and antibacterial activity were seen in an in vitro experiment utilising macrophages generated from mice bone marrow and treated with high glucose. Reduced phagocytosis was seen in peritoneal macrophages from diabetic mice in the same investigation [39]. This might be connected to macrophages' decreased glycolytic reserve and capacity as a result of their long-term sensitivity to high glucose levels.

Phagocytosis and adhesion capacity in RPMs of db/db mice decreases significantly thereby employing resident peritoneal macrophages (RPMs) obtained from mice. Additionally, compared to control mice, db/db mice showed enhanced macrophage polarisation shifting to M2 macrophages. A similar rise in M2 macrophage markers, such as Arginase 1 and IL-10, was observed in macrophages generated from mice bone marrow and exposed to high glucose for an extended length of time [40]. The immune response to bacterial infection may be weakened by this shifting since M2 macrophages have a low potential for microbicidal activity.

7. Ineffective natural killer cells

Innate immune responses to pathogens are mostly mediated by natural killer cells, and during the past 10 years, research into the protective effects of these cells against intracellular bacterial infections has acquired significant impetus. Numerous inhibitory and activating receptors control natural killer cells. Isolated NK cells from T2D patients were used to demonstrate the dysfunction of natural killer (NK) cells, which are crucial for containing invasive pathogens [41]. It was discovered that defects in the NK cell-activating receptors NKG2D and NKp46 were linked to functional defects in NK degranulation capacity [42].

NK cells with T-cell receptors are a special subset known as natural killer T (NKT) cells. They have the capacity to enhance a variety of immune responses and react to glycolipid antigens rather than peptide antigens. There is proof that NKT cells aid in host defence during *M. tuberculosis* infection by suppressing intracellular bacterial growth through cytolytic processes, promoting antigen-presenting cell (APC) maturation and activation, and modifying the sort of immune response elicited [43]. In experimental models of diabetes, the role of NKT cells in adipose tissue inflammation and glucose intolerance has been discussed. Increases in NKT cell numbers are seen in tuberculosis patients, and those with co-morbid diabetes had higher levels of NKT cells in their blood and bronchoalveolar lavage than those without TB [44]. This has been proposed as a helpful marker for active tuberculosis and may be a direct result of the elevated bacillary burden seen in these patients.

7.1 Impaired immune and complement system

In a study on rats the malfunction of complement activation was noted. They showed that elevated blood sugar levels were linked to a reduction in C4-fragment opsonization, which blocks the classical or lectin pathways of complement activation [44]. **Table 1** provides an overview of the potential pathways that lead to infection susceptibility in diabetics.

The results described in the previous section imply that islet macrophages have protective effects and help to maintain islet homeostasis. However, recent studies have also demonstrated that they play a significant role in the islet pathology in T2D [63]. The number of macrophages within islets was found to be increased in pancreas sections from T2D patients, C57BL/6 mice given a high-fat diet, db/db mice, and Goto-Kakizaki (GK) rats by immunohistochemical analysis [64]. Additionally, it has been claimed that high glucose or palmitate caused the production of chemokines from the islets, which aided in monocyte and neutrophil migration. This shows that the type 2 diabetic environment may encourage macrophage infiltration into pancreatic islets and stimulate chemokine production [65].

The build-up of macrophages within T2D islets points to their pathological function. It has been noted that macrophages perform seemingly incompatible tasks in islets as well as in other organs and conditions. In reality, recent research has shown that macrophages are actually extremely diverse [66]. According to in vitro research, Th1 cytokines alone or in combination with microbial products cause macrophages to activate in the traditional M1 manner, whereas Th2 cytokines (IL-4 and IL-13) cause an alternative type of activation known as M2 [53]. ActivatedM2-type macrophages enhance wound healing and may also modify immunological responses, whereas classically activatedM1-type macrophages play a key role in host defence by secreting proinflammatory cytokines and ROS. However, the phrase "M2 activation" is somewhat amorphous and is used to refer to a variety of M1-independent macrophage activation mechanisms [54].

M2 macrophages may therefore act differently depending on the environment. The functions of various macrophage subsets in the onset and development of disease, as well as their potential contributions to the preservation of homeostasis, are now well understood. Because macrophages have a variety of activation phenotypes, we examined the polarity of macrophage activation inside islets. We identified two distinct subpopulations of islets using flow cytometry: CD11b+Ly-6C+CD11b+F4/80+Ly-6C- T2D and CD11b+Ly-6C+CD11b+Ly-6C. healthier islet. CD11bhighF4/80−/+Ly-6C+ Diabetes type 2 and islet macrophage polarity (T2D). Healthy islets have a high percentage of resident macrophages that display CD11b+F4/80+Ly-6C [67].

Effects on immune system	Sources involved	Mechanism of action	References
Vanquished cytokine production	PBMCs isolated from healthy individuals	TGF-mediated reduction of IL-6, IL-2, and IL-10 production by PBMC; stimulation of cellular TGF- synthesis to inhibit mononuclear cell proliferation.	[45]
	PBMCs isolated from healthy individuals	Reduced IL-17A, which impairs immunological responses; Reduced IL-6 expression in CD14+ and CD16+ intermediate monocytes	[46]
	High fat diet-induced *db/db* obese mice	Suppressed IL-22 in blood plasma	[47]
	PBMCs isolated from healthy individuals and THP-1 human monocyte cell line	Defective type 1 IFN production	[48]
	T2D patients and healthy donors	Deficient glutathione accompanied with decreased production of IL-12 and IFγ	[49]
Disturbance in leukocyte recruitment	Mice treated with Streptozotocin- (C57BL/6)	Production of cytokines like CXCL1, CXCL2, IL-1, and TNF- is impaired	[50]
	C57BL/6 J (Wild Type) and C57BL/6 J (*db/db*) mice	Decreased expression of CAM causes a reduction in leukocyte movement, particularly cytotoxic CD8+ T cell migration.	[51]
Defective mechanism for recognition of pathogen	Mice treated with Streptozotocin- (C57BL/6)	Downregulated expression of TIRAP and TLR	[52]
Dysfunctional Neutrophil	Neutrophils isolated from T2D subjects	Increased resistin reduces the production of ROS in neutrophils	[53]
	Neutrophils isolated from healthy subjects	Inhibition of G6PD leading to production of impaired O_2^-	[54]
	Neutrophils isolated from healthy subjects	Coagulation and degranulation defect in neutrophil	[55]
	Neutrophils isolated from healthy subjects and T2D patients	Impaired neutrophil NET formation	[56]
	Neutrophils isolated from healthy subjects	Structural changes in C3b due to phagocytosis dysfunction of neutrophils in *S. aureus*	[57]

Effects on immune system	Sources involved	Mechanism of action	References
Dysfunction of macrophages and monocytes	Isolation of resident peritoneal macrophages from *db/db* mice and littermate controls (C57BL/6 J)	Impaired adhesion capacity and chemotaxis and in RPMs	[58]
		Expression of M2 phenotypes with anti-inflammatory properties	[59]
	Mice (C57BL/6 J) treated with streptozotocin and bone marrow-derived macrophages	Expression of M2 phenotypes with anti-inflammatory properties	[60]
		Decreased glycolytic reserve in macrophages' following prolonged exposure to high glucose	[61]
	PBMCs isolated from healthy individuals and T2D subjects	Supressed expression of Fcγ receptors on DM2 monocytes	[42]
Non-functional NK cell	PBMC isolated from T2D subjects	NK cell-activating receptor NKG2D and NKp46 abnormalities increase susceptibility to infections and cancer.	[40]
Antibody and complement effector inhibition	Streptozotocin-treated Wistar rat peritoneal cells	Impaired C4-fragment opsonization in hyperglycemic circumstances and suppression of complement activation through traditional or lectin pathways	[62]

Table 1.
The immunological mechanism underlying infection susceptibility in diabetics.

Monocytes/macrophages and CD11b+Ly-6Cmacrophages accumulate in T2D islets. T2D islet CD11b high F4/80+Ly 6C+monocytes/macrophages are also present. Islet-resident macrophages were predominately CD11b+Ly-6C cells with an M2-type phenotype under baseline circumstances. In comparison to control db/+ and KKTa animals, fractions of these M2-type cells were not altered in db/db or KKAy model T2D mice, respectively [55]. On the other hand, the T2D models had a specifically higher number of CD11b+Ly-6C+macrophages. These cells have an M1-type phenotype and express pro-inflammatory cytokines like IL-1 and TNF. As a result, in T2D islets, macrophage polarity seems to have switched toward M1 [56].

Inflammasome activation and High Glucose in T2D Islets. The polarity of macrophages within T2D islets is altered toward M1. Recent research has uncovered a number of mechanisms, including immune cell recruitment and the elevation of inflammatory cytokines (such IL-1), that underlie the activation of inflammatory processes within islets [57]. For instance, as was already established, the environment of type 2 diabetes may stimulate the creation of chemokines that encourage macrophage infiltration into pancreatic islets. Multiprotein complexes called inflammasomes are crucial for the development and release of IL-1. Two stimuli are necessary to initiate IL-1 secretion: the first stimulates pro-IL-1 protein expression, and the second activates inflammasomes, which in turn activate caspase I to cleave pro-IL-1 and produce mature IL-1 [58].

According to a recent study, minimally modified LDL, which triggers TLR4 signalling in macrophages and primes them to process Il-1, is one of the initial stimuli in T2D islets. The second stimulus was determined to be islet amyloid polypeptide (IAPP), a distinct polypeptide component of amyloid present in pancreatic islets and which is produced from cells in response to high glucose. IAPP, a soluble oligomer, activates the NLRP3 inflammasome and causes the islet macrophages to secrete IL-1 [59]. As a result, the interaction between macrophages and -cells is crucial for inflammasome activation within islets. Additionally, it has been demonstrated that high glucose-induced ROS production in cells causes the activation of the NLRP3 inflammasome and the release of IL-1 [60].

Islet inflammation leads to Cell dysfunction in T2D. Despite the fact that it seems as though high glucose levels are a necessary trigger for islet inflammation, a recent examination of -cell function revealed that impaired glucose tolerance was already present before glucose-induced insulin secretion deteriorated. This shows that cell dysfunction can start and progress in response to triggers other than excessive glucose levels [45]. FFAs are a potential contender for such stimulation. Clinical studies have shown that FFA levels are a reliable predictor of future T2D and a high consumption of saturated fatty acids has been associated with an increased risk of T2D. The most prevalent saturated FFA in blood is palmitate, and studies have shown that it has harmful effects on -cells that are collectively known as "lipotoxicity" [46].

Studies conducted in vitro have demonstrated that palmitate directly induces -cell lipotoxicity, at least in part through mechanisms predominantly involving ER stress and ROS. We created a technique to elevate non-esterified palmitate levels in the serum by injecting emulsified ethyl palmitate in order to assess the effects of palmitate on -cells in vivo. This model demonstrated that palmitate activates TLR4 to generate chemokines, such as CCL2 and CXC11, in cells [47]. These chemokines caused CD11b+Ly-6C+M1-type monocytes and macrophages to be attracted to the islets. Additionally, palmitate-induced -cell dysfunction was decreased when M1-type cells were prevented from accumulating by utilisingclodronate liposomes, demonstrating their causative involvement [48].

Additionally, it was discovered that M1 macrophages' production of proinflammatory cytokines, such as IL-1 and TNF-, encourages -cell dysfunction and that the vicious loop created by the secretion of chemokines by -cells and cytokines by M1 macrophages speeds up islet inflammation [49]. Similar to this, M1 macrophage increase within islets in T2D models (db/db and KKAy animals) appears to lead to cell dysfunction. These findings unequivocally demonstrate that inflammation-related islet dysfunction comes from the stimulation of inflammatory mechanisms [68].

Inflammation as a Pharmacological Target for T2D: Due to the role that IL-1 plays in the emergence of T2D and -cell dysfunction, therapeutic approaches that target the IL-1 receptor and IL-1 ligand have been developed. Rheumatoid arthritis is treated with recombinant human IL-1RA (anakinra), a medication that blocks IL-1 receptor signalling [69]. IL-1RA presumably inhibits both IL-1 and IL-1 signalling since it suppresses the IL-1 receptor. Anakinra was tested in a clinical trial to see if it could improve -cell function and glycaemic control in T2D patients. The anakinra group demonstrated improved HbA1c levels and serum C-peptide concentrations during oral glucose tolerance test (OGTT) with no significant differences in insulin sensitivity, indicating that improved cell function played the major role in the improvement in glucose tolerance [70].

A follow-up investigation showed that the decreased proinsulin-to-insulin ratio persisted for 39 weeks after the end of the treatment. In GK rats and mice fed a high-fat diet, the processes underlying these observations were further examined. IL-1RA enhanced insulin sensitivity and beta-cell activity in these animals by suppressing inflammation in insulin target tissues and islets. There was a noticeable decrease in islet macrophage counts in GK rats treated with IL-1RA, indicating that islet macrophages may be one of the targets of the anakinra treatment [71]. Finally, despite the drug improving the glucose disposition index during OGTT, no appreciable change in insulin sensitivity or -cell function was seen in a recent investigation assessing the effects of anakinra on obese adult individuals without T2D.

Additionally, IL-1-specific antibodies have been created. The therapeutic effects of gevok-izumab, a recombinant humanised monoclonal antibody that neutralises IL-1, were examined in T2D participants. The fact that this medication preserves IL-1 signalling and has a longer half-life (22–25 days), which lowers the frequency of administration and lowers the cost, may make it superior to anakinra [50]. Gevokizumab significantly decreased HbA1c, C-peptide secretion, and CRP at a low dose (0.3 mg/kg), but not at a high dose (0.03–0.1 mg/kg). The discovery that a large dose failed to exhibit positive effects may support the idea that IL-1 at low concentrations is advantageous for cells [51]. The scientists came to the conclusion that the right dosage and duration of gevokizumab therapy are essential for changing the immune system in T2D patients. Salsalate, a salicylic acid prodrug having inhibitory effects on the NF-B pathway, and TNF-inhibitors have both been investigated in T2D [72].

These research' encouraging findings are in line with the idea that T2D can be treated by reducing the inflammation induced by diabetes.

8. Future perspective

A serious global concern is the dual burden of intracellular bacterial infections and diabetes. The majority of current diagnostic and therapeutic research involves non-diabetic mice, and it is unclear whether these findings can be applied to people with diabetes given the obvious disparities in immune responses and disease mechanisms. There is considerable clinical and experimental evidence that a delay in innate immune system inflammatory signals is followed by delayed development of effective protective responses against intracellular bacterial infections, notwithstanding the complexity of the underlying immunopathology of diabetes. It is likely that a more multifaceted therapeutic approach will be required to address the complicated immunopathogenesis underlying diabetes, even while better glucose management may assist patients with intracellular infections and co-morbid diabetes. It will be easier to treat and manage disease in sensitive populations if we are aware of the mechanisms driving co-morbidities like diabetes, which profoundly affect the development of intracellular bacterial infections. Innovative, cost-effective strategies are desperately needed, especially in low- and middle-income nations where there has never been a greater convergence of non-communicable and communicable diseases. A multidisciplinary approach with an extensive study is required to tackle the present and future issues of the rising double burden of co-morbid intracellular bacterial infections leading to continued and widespread existence of non-communicable illnesses.

9. Conclusion

Diabetes is a metabolic disorder brought on by inflammation in an advanced immune system. Insulin resistance results in a multitude of immunological reactions that aggravate the inflammatory state and lead to hyperglycemia as a result of the inhibition of insulin signalling. Both issues with the innate immune response (including neutrophil and macrophage dysfunction) and deficiencies in the adaptive immune response are thought to contribute to the immune system's degradation against invasive infections in diabetics (including T cells). A deeper understanding of the processes of hyperglycemia that impair host defence against pathogens is crucial for the development of cutting-edge treatments to treat infections in diabetic patients and improve treatment outcomes.

Author details

Pratima Tripathi
Department of Biotechnology, National Institute of Pharmaceutical Education and Research, Lucknow, UP, India

*Address all correspondence to: pratimatripathi.lko@gmail.com

References

[1] Zimmet PZ, Magliano DJ, Herman WH, Shaw JE. Diabetes: A 21st century challenge. The Lancet Diabetes and Endocrinology. 2014;**2**(1):56-64. DOI: 10.1016/S2213-8587(13)70112-8

[2] Whiting DR, Guariguata L, Weil C, Shaw J. IDF diabetes atlas: Global estimates of the prevalence of diabetes for 2011 and 2030. Diabetes Research and Clinical Practice. 2011;**94**(3):311-321. DOI: 10.1016/J.DIABRES.2011.10.029

[3] Berbudi A, Rahmadika N, Tjahjadi AI, Ruslami R. Type 2 diabetes and its impact on the immune system. Current Diabetes Reviews. 2019;**16**(5):442-449. DOI: 10.2174/1573399815666191024085838

[4] Harding JL, Pavkov ME, Magliano DJ, Shaw JE, Gregg EW. Global trends in diabetes complications: A review of current evidence. Diabetologia. Jan 2019;**62**(1):3-16. DOI: 10.1007/s00125-018-4711-2. Epub 31 Aug 2018. PMID: 30171279

[5] Cerf ME. Beta cell dysfunction and insulin resistance. Front Endocrinol (Lausanne). 27 Mar 2013;**4**:37. DOI: 10.3389/fendo.2013.00037. PMID: 23542897; PMCID: PMC3608918

[6] Masini M, Bugliani M, Lupi R, del Guerra S, Boggi U, Filipponi F, et al. Autophagy in human type 2 diabetes pancreatic beta cells. Diabetologia. Jun 2009;**52**(6):1083-1086. 2009. DOI: 10.1007/s00125-009-1347-2. Epub 15 Apr 2009. PMID: 19367387

[7] Berbudi A, Rahmadika N, Tjahjadi AI, Ruslami R. SCIENCE BENTHAM send orders for reprints to reprints@benthamscience.net type 2 diabetes and its impact on the immune system. Current Diabetes Reviews. 2020;**16**:442-449. DOI: 10.2174/1573399815666191024085838

[8] Parry SA, Smith JR, Corbett TRB, Woods RM, Hulston CJ. Short-term, high-fat overfeeding impairs glycaemic control but does not alter gut hormone responses to a mixed meal tolerance test in healthy, normal-weight individuals*. British Journal of Nutrition. 2017;**117**:48-55. DOI: 10.1017/S0007114516004475

[9] Carmichael L, Keske MA, Betik AC, et al. Is vascular insulin resistance an early step in diet-induced whole-body insulin resistance? Nutrition & Diabetes. Jun 2022;**12**(1):31-43. DOI: 10.1038/s41387-022-00209-z. PMID: 35676248; PMCID: PMC9177754

[10] Zhang A, Li M, Qiu J, Sun J, Su Y, Cai S, et al. The relationship between urinary albumin to creatinine ratio and all-cause mortality in the elderly population in the Chinese community: A 10-year follow-up study. BMC Nephrology. Jan 2022;**23**(1):16-26. DOI: 10.1186/s12882-021-02644-z. PMID: 34983421; PMCID: PMC8729014

[11] Lawal B, Sani S, Onikanni AS, Ibrahim YO, Agboola AR, Lukman HY, et al. Preclinical anti-inflammatory and antioxidant effects of Azanza garckeana in STZ-induced glycemic-impaired rats, and pharmacoinformatics of it major phytoconstituents. Biomedicine & Pharmacotherapy. 2022;**152**:113196. DOI: 10.1016/j.biopha.2022.113196

[12] Tomita T. Apoptosis in pancreatic β-islet cells in Type 2 diabetes. Bosnian Journal of Basic Medical Sciences. 2016;**16**(3):162-179. DOI: 10.17305/bjbms.2016.919

[13] Waddell HMM, Zhang JZ, Hoeksema KJ, McLachlan JJ, McLay JC,

Jones PP. Oxidation of RyR2 has a biphasic effect on the threshold for store overload-induced calcium release. Biophysical Journal. 2016;**110**(11):2386-2396. DOI: 10.1016/j.bpj.2016.04.036

[14] He Q, Xu JY, Gu J, Tong X, Wan Z, Gu Y, et al. Piperine is capable of improving pancreatic β-cell apoptosis in high fat diet and streptozotocin induced diabetic mice. Journal of Functional Foods. 2022;**88**:104890. DOI: 10.1016/j.jff.2021.104890

[15] Oguntibeju OO, Aboua GY, Omodanisi EI. Effects of Moringa oleifera on oxidative stress, apoptotic and inflammatory biomarkers in streptozotocin-induced diabetic animal model. South African Journal of Botany. 2020;**129**:354-365. DOI: 10.1016/j.sajb.2019.08.039

[16] Lee JH, Mellado-Gil JM, Bahn YJ, Pathy SM, Zhang YE, Rane SG. Protection from β-cell apoptosis by inhibition of TGF-β/Smad3 signaling. Cell Death and Disease 11, 2020:184-199. DOI: 10.1038/s41419-020-2365-8

[17] Schuetz P, Castro P, Shapiro NI. Diabetes and Sepsis: Preclinical findings and clinical relevance. Diabetes Care. Mar 2011;**34**(3):771-778. DOI: 10.2337/dc10-1185. PMID: 21357364; PMCID: PMC3041224

[18] Esper AM, Moss M, Martin GS. The effect of diabetes mellitus on organ dysfunction with sepsis: an epidemiological study. Crit Care. 2009;**13**(1):R18-23. DOI: 10.1186/cc7717. Epub 13 Feb 2009. PMID: 19216780; PMCID: PMC2688136

[19] Gupta S, Maratha A, Siednienko J, Natarajan A, Gajanayake T, Hoashi S, et al. Analysis of inflammatory cytokine and TLR expression levels in type 2 diabetes with complications. Scientific Reports. 9 Aug 2017;7(1):7633-7643. DOI: 10.1038/s41598-017-07230-8. Erratum in: Sci Rep. 5 Apr 2018;**8**(1):5768. PMID: 28794498; PMCID: PMC5550417

[20] Neely AN, Clendening CE, Gardner J, Greenhalgh DG. Gelatinase activities in wounds of healing-impaired mice versus wounds of non-healing-impaired mice. Journal of Burn Care and Rehabilitation. 2000;**21**(5):395-402. DOI: 10.1097/00004630-200021050-00001

[21] Berrou J, Fougeray S, Venot M, Chardiny V, Gautier J-F. Natural killer cell function, an important target for infection and tumor protection, is impaired in type 2 diabetes. PLoS One. 2013;**8**(4):62418. DOI: 10.1371/journal.pone.0062418

[22] Hiriart M, Hamasaki H, Clinic H, Villa AR, Marichal-Cancino BA, Chávez-Reyes J, et al. Susceptibility for some infectious diseases in patients with diabetes: The key role of Glycemia. Frontiers in Public Health | Www.Frontiersin.Org. 2021;**9**:559595. DOI: 10.3389/fpubh.2021.559595

[23] Karima M, Kantarci A, Ohira T, Hasturk H, Jones VL, Nam BH, et al. Enhanced superoxide release and elevated protein kinase C activity in neutrophils from diabetic patients: association with periodontitis. Journal of Leukocyte Biology. Oct 2005;**78**(4):862-870. DOI: 10.1189/jlb.1004583. Epub 2005 Aug 4. PMID: 16081595; PMCID: PMC1249507

[24] Martinez PJ, Mathews C, Actor JK, Hwang S-A, Brown EL, De Santiago HK, et al. Impaired CD4+ and T-helper 17 cell memory response to Streptococcus pneumoniae is associated with elevated glucose and percent glycated hemoglobin A1c in Mexican Americans with type 2 diabetes

mellitus. Translational Research. 2014;**163**(1):53-63. DOI: 10.1016/j.trsl.2013.07.005

[25] Tanji N, Markowitz GS, Fu C, Kislinger T, Taguchi A, Pischetsrieder M, et al. Expression of advanced glycation end products and their cellular receptor RAGE diabetic nephropathy and nondiabetic renal disease. Journal of the American Society of Nephrology. 2000;**11**(9):1656-1666. DOI: 10.1681/ASN.V1191656

[26] Holland SM, Gallin JI, Mandell, Douglas. Evaluation of the patient with suspected immunodeficiency. In: Bennett's Principles and Practice of Infectious Diseases. 7th ed. Vol. 12. USA: Churchill Livingstone; 2010. pp. 167-78

[27] Jalkanen S, Salmi M. Lymphocyte adhesion and trafficking. Clinical Immunology. 2019;Chapter 11:171-182.e1. First edition Part 1. DOI: 10.1016/B978-0-7020-6896-6.00011-9

[28] Gorain B, Bhattamishra SK, Choudhury H, Nandi U, Pandey M, Kesharwani P. Overexpressed receptors and proteins in lung Cancer. In: Nanotechnology-Based Targeted Drug Delivery Systems for Lung Cancer. United Kingdom: Woodhead Publishing; 2019. pp. 39-75. DOI: 10.1016/B978-0-12-815720-6.00003-4

[29] Shah B, Burg N, Pillinger MH. Neutrophils. In: Kelley and Firestein's Textbook of Rheumatology. Baishideng Publishing Group; 2017. pp. 169-188.e3. DOI: 10.1016/B978-0-323-31696-5.00011-5

[30] Hyde DM, Simon SI. Inflammatory cells of the lung: Polymorphonuclear leukocytes. In: Comprehensive Toxicology. 2nd ed. Vol. 8. Elsevier; 2010. pp. 115-127. DOI: 10.1016/B978-0-08-046884-6.00906-4

[31] Roep BO, Thomaidou S, van Tienhoven R, Zaldumbide A. Type 1 diabetes mellitus as a disease of the β-cell (do not blame the immune system?). Nature Reviews Endocrinology. 2021;**17**:150-161. DOI: 10.1038/s41574-020-00443-4

[32] Pino SC, Kruger AJ, Bortell R. The role of innate immune pathways in type 1 diabetes pathogenesis. Current Opinion in Endocrinology, Diabetes and Obesity. Apr 2010;**17**(2):126-30. DOI: 10.1097/MED.0b013e3283372819. PMID: 20125005; PMCID: PMC2905794

[33] Scaglione R, Elisei R, Marchetti P. International Journal of Molecular Sciences. 2022;**2022**:3657. DOI: 10.3390/ijms23073657

[34] Steiniger B, Barth P, Hellinger A. The perifollicular and marginal zones of the human splenic white pulp: Do fibroblasts guide lymphocyte immigration? American Journal of Pathology. 2001;**159**(2):501-512. DOI: 10.1016/S0002-9440(10)61722-1

[35] Martins M, Boavida JM, Raposo JF, Froes F, Nunes B, Ribeiro RT, et al. Diabetes hinders community-acquired pneumonia outcomes in hospitalized patients. BMJ Open Diabetes Research and Care. 2016;**4**:e000181. DOI: 10.1136/bmjdrc-2015-000181

[36] Jafar N, Edriss H, Nugent K. The effect of short-term hyperglycemia on the innate immune system. The American Journal of the Medical Sciences. 2016;**351**(2):201-211. DOI: 10.1016/J.AMJMS.2015.11.011

[37] Stojanovi SD, Fiedler J, Bauersachs J, Thum T, Sedding DG. Senescence-induced inflammation: An important player and key therapeutic target in atherosclerosis. European Heart Journal. 14 Aug 2020;**41**(31):2983-2996. DOI: 10.1093/

eurheartj/ehz919. PMID: 31898722; PMCID: PMC7453834

[38] Feng X, Chen W, Ni X, Little PJ, Xu S, Tang L, et al. Metformin, macrophage dysfunction and atherosclerosis. Frontiers in Immunology. 7 Jun 2021;**12**:682853. DOI: 10.3389/fimmu.2021.682853. PMID: 34163481; PMCID: PMC8215340

[39] Sukhorukov VN, Khotina VA, Chegodaev YS, Ivanova E, Sobenin IA, Orekhov AN. Lipid metabolism in macrophages: Focus on Atherosclerosis. Biomedicine. 1 Aug 2020;**8**(8):262-277. DOI: 10.3390/biomedicines8080262. PMID: 32752275; PMCID: PMC7459513

[40] Marchio P, Guerra-Ojeda S, Vila JM, Aldasoro M, Victor VM, Mauricio MD. Targeting early atherosclerosis: A focus on oxidative stress and inflammation. Oxidative Medicine and Cellular Longevity. 1 Jul 2019;**2019**:8563845. DOI: 10.1155/2019/8563845. PMID: 31354915; PMCID: PMC6636482

[41] Perera Molligoda Arachchige AS. Human NK cells: From development to effector functions. innate immune. Apr 2021;**27**(3):212-229. DOI: 10.1177/17534259211001512. Epub 24 Mar 2021. PMID: 33761782; PMCID: PMC8054151

[42] Waggoner SN, Von Herrath MG, Penaloza-Macmaster P, Lang PA, Crome SQ, Xu HC, et al. NK cells regulate CD8 + T cell mediated autoimmunity. Frontiers in Cellular and Infection Microbiology. 2020;**10**:36. DOI: 10.3389/fcimb.2020.00036. www.Frontiersin.Org

[43] O'shea D, Hogan AE. Dysregulation of Natural Killer Cells in Obesity. Cancers (Basel). 23 Apr 2019;**11**(4):573-585. DOI: 10.3390/cancers11040573. PMID: 31018563; PMCID: PMC6521109

[44] Gardner G, Fraker CA. Natural killer cells as key mediators in type I diabetes immunopathology. Frontiers in Immunology. 20 Aug 2021;**12**:722979. DOI: 10.3389/fimmu.2021.722979. PMID: 34489972; PMCID: PMC8417893

[45] Harris CL, Pouw RB, Kavanagh D, Sun R, Ricklin D. Developments in anti-complement therapy; from disease to clinical trial. Molecular Immunology. 2018;**102**:89-119. DOI: 10.1016/J.MOLIMM.2018.06.008

[46] Kulak K, Westermark GT, Papac-Milicevic N, Renström E, Blom AM, King BC, et al. The human serum protein C4b-binding protein inhibits pancreatic IAPP-induced inflammasome activation. Diabetologia. Aug 2017;**60**(8):1522-1533. DOI: 10.1007/s00125-017-4286-3. Epub 12 May 2017. PMID: 28500395; PMCID: PMC5491568

[47] Nagaraj V, King B, Storm P, Vikman P, Ottosson-Laakso E, Blom AM, et al. Complement inhibitor CD55 governs the integrity of membrane rafts in pancreatic beta cells, but plays no role in insulin secretion. Biochemical and Biophysical Research Communications. 2015;**460**(3):518-524. DOI: 10.1016/J.BBRC.2015.03.062

[48] Dos Santos RS, Marroqui L, Grieco FA, Marselli L, Suleiman M, Henz SR, et al. Protective role of complement C3 against cytokine-mediated b-cell apoptosis. Endocrinology. 2017;**158**:2503-2521. DOI: 10.1210/en.2017-00104

[49] Golec E, Rosberg R, Zhang E, Renström E, Renström R, Blom AM, et al. A cryptic non-GPI-anchored cytosolic isoform of CD59 controls insulin exocytosis in pancreatic b-cells by interaction with SNARE proteins. FASEB J. Nov 2019;**33**(11):12425-12434. DOI: 10.1096/fj.201901007R. Epub 14 Aug 2019. PMID: 31412214; PMCID: PMC6902737

[50] Lo JC, Ljubicic S, Leibiger B, Kern M, Leibiger IB, Moede T, et al. Adipsin is an Adipokine that improves β cell function in diabetes. Cell. 2014;**158**(1):41-53. DOI: 10.1016/J.CELL.2014.06.005

[51] Lee MJ, Wu Y, Fried SK. Adipose tissue heterogeneity: Implication of depot differences in adipose tissue for obesity complications. Molecular Aspects of Medicine. 2013;**34**(1):1-11. DOI: 10.1016/J.MAM.2012.10.001

[52] Li R, Coulthard LG, Wu MCL, Taylor SM, Woodruff TM. C5L2: A controversial receptor of complement anaphylatoxin, C5a. FASEB J. Mar 2013;**27**(3):855-864. DOI: 10.1096/fj.12-220509. Epub 13 Dec 2012. PMID: 23239822

[53] Ahlqvist E, Storm P, Käräjämäki A, Martinell M, Dorkhan M, Carlsson A, et al. Novel subgroups of adult-onset diabetes and their association with outcomes: A data-driven cluster analysis of six variables. The Lancet Diabetes and Endocrinology. 2018;**6**(5):361-369. DOI: 10.1016/S2213-8587(18)30051-2

[54] Kolev M, Kemper C. Keeping It All Going-Complement Meets Metabolism. Frontiers in Immunology. 2017;**8**:1-18. DOI: 10.3389/fimmu.2017.00001. PMID: 28149297; PMCID: PMC5241319

[55] Heeger PS, Kemper C. Novel roles of complement in T effector cell regulation. Immunobiology. 2012;**217**(2):216-224. DOI: 10.1016/J.IMBIO.2011.06.004

[56] Carroll MC, Isenman DE. Regulation of humoral immunity by complement. Immunity. 2012;**37**(2):199-207. DOI: 10.1016/J.Immuni.2012.08.002

[57] Pouw RB, Brouwer MC, de Gast M, van Beek AE, van den Heuvel LP, Schmidt CQ, et al. Potentiation of complement regulator factor H protects human endothelial cells from complement attack in aHUS sera. Blood Advances. 26 Feb 2019;**3**(4):621-632. DOI: 10.1182/bloodadvances.2018025692. PMID: 30804016; PMCID: PMC6391659

[58] de Cordoba SR, Tortajada A, Harris CL, Morgan BP. Complement dysregulation and disease: From genes and proteins to diagnostics and drugs. Immunobiology. 2012;**217**(11):1034-1046. DOI: 10.1016/J.IMBIO.2012.07.021

[59] Ricklin D, Reis ES, Lambris JD. Complement in disease: A defence system turning offensive. Nature Reviews Nephrology. Jul 2016;**12**(7):383-401. DOI: 10.1038/nrneph.2016.70. Epub 2016 May 23. PMID: 27211870; PMCID: PMC4974115

[60] Marc Y. Donath, Inflammation as a sensor of metabolic stress in obesity and type 2 diabetes. Endocrinology. 1 Nov 2011;**152**(11):4005-4006. DOI: 10.1210/en.2011-1691

[61] Shim K, Begum R, Yang C, Wang H. Complement activation in obesity, insulin resistance, and type 2 diabetes mellitus conflict-of-interest statement. World Journal of Diabetes. 2020;**11**(1):1-12. DOI: 10.4239/wjd.v11.i1.1

[62] Lascar N, Brown J, Pattison H, Barnett AH, Bailey CJ, Bellary S. Type 2 diabetes in adolescents and young adults. The Lancet Diabetes & Endocrinology. 2018;**6**(1):69-80. DOI: 10.1016/S2213-8587(17)30186-9

[63] De Borst MH, Poppelaars F, Böhringer S, Valoti E, Noris M, Perna A, et al. Impact of a complement factor H gene variant on renal dysfunction, cardiovascular events, and response to ACE inhibitor therapy in type 2 diabetes. Article 681. Frontiers in

Genetics. 2019;**10**:681. DOI: 10.3389/fgene.2019.00681

[64] King BC, Blom AM. Complement in metabolic disease: Metaflammation and a two-edged sword. Seminars Immunopathology. 2021;**43**:829-841. DOI: 10.1007/s00281-021-00873-w

[65] Dennis JM, Shields BM, Henley WE, Jones AG, Hattersley AT. Disease progression and treatment response in data-driven subgroups of type 2 diabetes compared with models based on simple clinical features: An analysis using clinical trial data. The Lancet Diabetes & Endocrinology. 2019;**7**(6):442-451. DOI: 10.1016/S2213-8587(19)30087-7

[66] Zou X, Zhou X, Zhu Z, Ji L. Novel subgroups of patients with adult-onset diabetes in Chinese and US populations. Lancet Diabetes Endocrinology. Jan 2019;**7**(1):9-11. DOI: 10.1016/S2213-8587(18)30316-4. PMID: 30577891

[67] Han G, Geng S, Li Y, Chen G, Wang R, Li X, et al. cdT-cell function in sepsis is modulated by C5a receptor signalling. Immunology. 2011 Blackwell Publishing Ltd, Immunology;**133**:340-349. DOI: 10.1111/j.1365-2567.2011.03445.x

[68] Krus U, King BC, Nagaraj V, Gandasi NR, Sjölander J, Buda P, et al. The complement inhibitor CD59 regulates insulin secretion by modulating Exocytotic events. Cell Metabolism. 2014;**19**(5):883-890. DOI: 10.1016/J.CMET.2014.03.001

[69] King BC, Kulak K, Krus U, Rosberg R, Golec E, Wozniak K, et al. Complement component C3 is highly expressed in human pancreatic islets and prevents β cell death via ATG16L1 interaction and autophagy regulation. Cell Metabolism. 2019;**29**(1):202-210.e6. DOI: 10.1016/J.CMET.2018.09.009

[70] Sorbara MT, Foerster EG, Tsalikis J, Abdel-Nour M, Mangiapane J, Sirluck-Schroeder I, et al. Complement C3 drives autophagy-dependent restriction of Cyto-invasive Bacteria. Cell Host & Microbe. 2018;**23**(5):644-652.e5. DOI: 10.1016/J.CHOM.2018.04.008

[71] Kolev M, Dimeloe S, Le Friec G, Navarini A, Arbore G, Povoleri GA, et al. Complement regulates nutrient influx and metabolic reprogramming during Th1 cell responses. Immunity. 2015;**42**(6):1033-1047. DOI: 10.1016/J.IMMUNI.2015.05.024

[72] Lim J, Iyer A, Suen JY, Seow V, Reid RC, Brown L, et al. C5aR and C3aR antagonists each inhibit diet-induced obesity, metabolic dysfunction, and adipocyte and macrophage signaling. FASEB J. Feb 2013;**27**(2):822-831. DOI: 10.1096/fj.12-220582. Epub 1 Nov 2012. PMID: 23118029

Chapter 6

Perspective Chapter: Role of Immunosuppressive and Immunomodulatory Agents in Cancer

Poppy Diah Palupi, Mohammed Safwan Ali Khan and Nur Dina Amalina

Abstract

Immunosuppressants offer some benefits and disadvantages. Like a blade with two edges, immunosuppressants are categorized as drugs but also cause decreased immunity, which eventually cause cancer. Immunosuppressants are widely used in organ transplantation patients and autoimmune illnesses to suppress the immune response and provide a significant risk of cancer. According to epidemiological and cancer research, malignancies are higher among transplant patients. However, the risk varies significantly between studies due to methods and patient selection variations. A more accurate illustration of the effects of mild-to-moderate immunosuppression concerning the risk of cancer can be seen in the rising use of immunosuppressant medications in non-transplant patients. Generally, cancer cells have an approach to avoid immune surveillance and create a complex balance in which many immune subtypes may be responsible for controlling tumor development, metastasis, and resistance. Therefore, the main objective of most cancer immunotherapies is to reestablish effective immune control. Immunomodulators help to maintain immune system function and promote the immune system's capacity to fight and defeat cancer. One of them is immune checkpoint inhibitors.

Keywords: immune system, immunosuppressant, immunomodulator, immune checkpoint inhibitor, cancer

1. Introduction

Malignancies are reported to be linked with the immune suppression system. Consequently, approximately, about 8.2 million annual casualties are expected to increase [1]. The concept, innate and adaptive immune cells can regulate tumor growth. However, neoplasm tissue tumors are identifiable as malignant cells and evolve new defense mechanisms that imitate peripheral immunological tolerance to fight against tumoricidal strikes [2]. In addition, due to the cancerous cells having antigens that make them different from normal cells, the immune system can find cells that have become cancerous [3]. The immune system recognizes and eliminates abnormal cells as part of its normal function, most likely preventing or slowing cancer progression [4].

Present chapter discusses the significance of comprehending the immune system's function in the emergence of cancer, including the often prescribed immunosuppressant medications for autoimmune and organ transplant patients. Also, this manuscript shows the importance of the immunomodulators, including immune checkpoint blockade, in cancer immunotherapy.

2. Cancer and immune suppression

The host immune system is well established to contribute to the evolution and progression of cancer, as significant as the tumor immune system. The complex interactions between the immune system and the tumor commonly occur in either the tumor's immune deterrence or the termination of cancer [5]. Moreover, significant discoveries in the last few decades have demonstrated that the immune system plays an important role in maintaining the equivalence between immune recognition and cancer development. And it might both promote and inhibit tumor growth [6].

Immune cells that have entered the tumor microenvironment (TME) regulate the growth and dissemination of cancer (TME) [7]. The disruption of the TME induces an inflammatory immune response, as evidenced by the presence of innate and adaptive immune cells in histopathological examinations, and is classified as tumor growth [8]. Interactions between the morphological and molecular elements of the TME through a complex and multistep metastatic cascade enable cancer cells to spread from the initial site to distant regions and become invasive [9]. Immune evasion also frequently occurs due to interactions between the elements of the immune system and the tumor cells in the TME, which promotes the development of tumors [10].

Tumor formation is initiated when immune cells like CD8+ T cells and natural killer (NK) cells attack and destroy most cancer cells. However, specific tumor cells can avoid these immune defenses by suppressing effector cells or stimulating tolerogenic cells during the immunological escape period. External and internal mechanisms that reduce antitumor immune responses and increase immunosuppressive cells, like regulatory T cells (Treg) and myeloid-derived suppressor cells (MDSCs), promote immune surveillance escape [11]. Regulatory T (Treg) cells expressing Foxp3+ reduce the dysfunctional immune response to self-antigens and the antitumor immune response. Additionally, Treg cell infiltration into tumor tissues is frequently associated with poor clinical outcomes [12]. Autoimmune diseases occur when Treg cells are insufficient due to immune suppression of Foxp3+, CD25+, and CD4+ Treg cells; these cells have been identified as one of the most important mechanisms of immunological self-tolerance [13]. In addition, a study has shown that Treg cell ablation can induce antitumor immunity effectively. However, it can also result in autoimmunity, particularly if Treg cells are eliminated systemically [14].

Treg cells have the ability to regulate T-cells, B-cells, NK cells, dendritic cells (DCs), and macrophages via humoral and cell–cell contact pathways. Several molecules, including TGF (transforming growth factor), GITR (glucocorticoid-induced TNF receptor), CTLA-4 (cytotoxic T-lymphocyte-associated protein 4), LAG3 (lymphocyte-activation gene 3), IL-2, IL-10, IL-35, granzyme B, adenosine, and cAMP, are implicated in Treg-mediated suppression pathways [15]. Foxp3+ regulates the expression of these molecules, and deficiencies of IL-2, CD25, CD122, and CTLA-4 are associated with autoimmune disorders. According to these hypotheses, only a few molecules have their expressions directly or indirectly regulated by Foxp3+, such as IL-2, IL-2 receptor subunits, and CTLA-4. The absence of these molecules results in severe autoimmune disorders and the loss of Treg-suppressive function [16–18].

In addition, the suppression mechanisms dependent on Tregs are necessary for establishing self-tolerance and have a significant impact on tumor immunity. The expression of CD25 and CTLA-4, dependence on exogenous IL-2, and T-cell receptor (TCR) activation produce Treg functions, particularly suppression mediated by Tregs. Regarding tumor immunity, these molecular mechanisms are also effective targets for regulating the activity and expansion of Tregs [16]. Therefore, recent advances in cancer immunotherapy targeted Treg cells suggest that Treg depletion or functional modification may be facilitated by Treg-specific drugs. These molecules are CD25, GITR, OX-40, and LAG3 [12]. CTLA-4, programmed cell death receptor-1 (PD-1), and programmed cell death ligand-1 (PD-L1) will be discussed further in this chapter.

3. Cancer and immunosuppressive agents

Since their discovery, many immunosuppressive medications have been used in transplantation and autoimmune diseases. Many organ transplant recipients, for instance, take medication to suppress their immune systems and reduce the graft rejection episodes. The immunosuppressive therapy after transplantation aims to prevent acute and chronic rejection while reducing adverse pharmacological effects in transplant recipients. The majority of therapies modify immune response mechanisms but lack immunological specificity [19].

Any immunosuppressive medication carries a potentially serious risk of cancer. Due to the increasing use of immunosuppressive medications among transplant and non-transplant patients, it is possible to define the effects of mild-to-moderate immunosuppression on the risk of neoplasms [20]. The use of immunosuppressive drugs in organ transplant patients is associated with a wide range of adverse side effects. The potential cancer risks—well-documented since the late 1960s—represent a significant cause of morbidity, mortality, and late failure in patients who otherwise have healthy grafts. Since transplantation first became popular, two agents have been utilized: azathioprine and corticosteroids [21]. Tacrolimus, cyclosporine, mycophenolate mofetil, everolimus, and sirolimus are the most common immunosuppressive drugs used in the combination therapy [19].

Cancers in transplant patients have been the subject of the current study, collecting information from single and multicenter studies. The type of malignancies and estimated risks differ significantly from study to study due to a variety of factors, including geographic differences, the use of various immunosuppressive regimens and antiviral therapy prevention, the duration of follow-up, the type of organ transplant and multiple techniques for estimating the occurrence [20]. The malignancy incidence in transplant recipients is higher in young adults, with significant clinical aggressiveness and a relatively short time in initiation after transplantation. Additionally, a key risk factor for immunosuppressive medication use is the dosing frequency and schedule [22].

A high prevalence of cancer among transplant recipients was seen/observed. Still, there is a debate over which factors—such as the type of immunosuppressive regimens, the overall level of immunosuppression, the course of treatment, or the dosage—is most important to assess the risks involved. The early agent used, azathioprine, can potentially cause cancer directly or indirectly. A study in premalignant dysplastic keratotic lesions showed that azathioprine might have a carcinogenic effect rather than simply suppressing the immune system [23]. Although cyclosporine unexpectedly showed the high rates of lymphomas and Kaposi's sarcomas, there is no substantial evidence that this medication increases the risk of tumors compared to those seen with the other immunosuppressive drugs such as, traditional azathioprine-based regimens [24]. On the other hand, tacrolimus-induced post-transplant malignancies were shown to have a high prevalence and pathological characteristics comparable to other immunosuppressive drugs [25].

In addition to treating non-transplant patients, such as Inflammatory Bowel Disease (IBD), immunomodulators such as thiopurines or methotrexate and TNF-antagonists may also reduce the incidence of inflammation-related cancers. However, although there is little chance of developing cancer when taking azathioprine in non-transplant patients, there is a potential risk that may rise with time and in a dose-dependent manner [26]. Although cyclophosphamide is typically used for cancer patients or bone marrow transplantation regimens, its immunosuppressive effects have also been applied to many chronic inflammatory diseases. For example, in people treated for cancer or non-malignant illnesses, it was shown to have the ability to increase the risk of bladder cancer. However, long-term cyclophosphamide therapy in a non-neoplasm patient is linked to an increased frequency of some malignancies, suggesting an immunosuppressive agent's potential side effect [27].

Immunomodulators and biological agents affect the immune system and might promote cancer development [28]. Thiopurines and methotrexate contribute to the development of the cancer by activating oncogenes, altering DNA directly, reducing physiologic immunosurveillance of malignant cells, and impairing the immune system's capacity to regulate oncogenic viruses [27, 29]. Infliximab, a chimeric IgG antibody, is primarily directed against TNF-α to neutralize the cytotoxic effects in a dose-dependent manner and has recently been licensed to treat rheumatoid arthritis and Crohn's disease [20]. However, TNF-α has many different impacts on the immune system, but the carcinogenic potential is less understood because of unreliable molecular information. TNF-α has been demonstrated to have antitumor activity by inducing the cellular death of malignant cells. Moreover, as a pro-tumor inflammatory cytokine, TNF-α is generated by most tumors to stimulate cellular survival and accelerate cancer growth [30, 31].

Another immunosuppressant agent that is used in the transplant population is Rapamycin. Streptomyces hygroscopicus is the source of the fermentation product rapamycin (RAPA), also referred to as sirolimus [32]. Many studies revealed that RAPA was a potent immunosuppressive drug to prevent allograft rejection in the heart, liver, lung, and kidney transplantation [33–35]. Sirolimus, and its analogs, including deforolimus, everolimus, and temsirolimus (a rapamycin prodrug), block the mechanistic target of Rapamycin (mTOR). Therefore, rapalogs are used in some clinical applications, such as organ transplant management and cancer therapy. The immunosuppressive properties of Rapalogs justified their use in organ recipients. Despite the fact that rapalogs were predicted to promote tumor growth and increase cancer incidence, they are frequently used in cancer treatment [36].

Commonly, growth factors, nutrient-rich environments, and oxygen levels excessively stimulate cultured cells. The environment stimulates growth-promoting pathways, including the PI3K/mTOR axis and mitogen-activated protein kinases (MAPKs). The activation of the oncosuppressor p53 and the accumulation of cell cycle inhibitors, such as cyclin-dependent kinase inhibitor 1A (CDKN1A) and cyclin-dependent kinase inhibitor 2A, can induce a cell cycle arrest under certain stress conditions (CDKN2A). In the case of malignant cells, growth factors and oncogenic signaling pathways continue to activate cultured cells by promoting mTOR and MAPK signaling even though the cell cycle has finished [37, 38]. Therefore, rapalogs are increasingly recommended for cancer treatment, especially for mTOR-dependent cancer subtypes [39, 40]. Thus, rapalogs could be thought of as anti-inflammatory substances that demonstrated anticancer. In addition, Rapamycin and its analogs also lowered the risk of cancer related to organ transplants and extended the overall and disease-free longevity of patients with certain malignancies.

4. Immune checkpoint inhibitors (ICI) as immunomodulatory agents

In adaptive immunity, two immune cells, B and T. B-cells, recognize circulating antigens in their natural state and produce protective antibodies in response [41]. T-cells are a powerful weapon the immune system uses to fight cancer [42]. T-cells identify peptide antigens from intracellularly degraded proteins filled onto the Cell's surface of *major histocompatibility complex (*MHC) molecules, and this process is known as antigen presentation [43]. Immunological checkpoints on the cell surface are activated when their surface proteins recognize and bind to partner proteins on other cells, such as specific tumor cells [42]. Self-tolerance requires immune checkpoints to prevent autoimmunity and protect tissues from destruction [44]. Tregs are drawn to cancer cells, which causes them to express less tumor antigen and release immune-suppressive cytokines that activate inhibitory immunological checkpoints [10] that create an immunosuppressive TME [45]. Immune checkpoint inhibitors work by obstructing particular inhibitory pathways' actions to combat immunosuppressive conditions [44]. The "brake system" of the immune system that cancers routinely exploit to halt immunological responses and defend themselves are immune checkpoints. Checkpoint inhibitors can generate new immune responses against cancer and strengthen already-existing ones to remove malignant cells.

The CTLA-4, PD-1, and PD-L1 are the common inhibitory checkpoints [46]. These antibodies have biological effects on different body parts during the T cell's lifecycle [47]. In addition, they functionally complement one another, establishing that T cell responses retain self-tolerance while defending the body from infections and cancer [43]. The immune checkpoint inhibitors approved by the FDA are shown in **Table 1**.

4.1 CTLA-4

In humans, CTLA4 is the first immune-checkpoint receptor to be studied; its located on T cells and controls T cell activation in the early stages of infection [48]. *CTLA-4* is a novel immunoglobulin superfamily member resembling CD28, structurally and pharmacologically [49]. CTLA-4 and CD28 are expressed exclusively in the hematopoietic compartment and found in the exact location of chromosome 2 (2q33.2). Furthermore, CTLA-4 and CD28 have the most sequence similarity in their extracellular binding domain; they bind to the identical CD80 and CD86 ligands expressed by antigen-presenting cells (APCs) [50]. Further characterization revealed that CD28 and CTLA4 have opposing immunoregulatory functions. CTLA4 inhibits T cell activation in a number of ways, including by directly opposing CD28, competing for co-stimulatory ligands, preventing the production of immunological conjugates,

Target	Drugs	Mechanism of action	Approval year	Treatment
CTLA-4	Ipilimumab	Inhibited CTLA-4, T-cell activation	2010	CRC, HCC, melanoma, mesothelioma, NSCLC, RCC
PD-1	Nivolumab	Inhibited PD-1, T-cell activation	2014	CRC, esophageal SCC, HCC, HL, HNSCC, melanoma, mesothelioma, NSCLC, RCC, UC
	Pembrolizumab	Inhibited PD-1, T-cell activation	2014	BC, CVC, CRC, CSCC, EnC, EsC, GC, HCC, HL, HNSCC, melanoma, mesothelioma, MCC, NSCLC, LBCL, RCC, SCLC, UC
	Cemiplimab	Inhibited PD-1, T-cell activation	2018	BCC, CSCC, NSCLC
	Dostarlimab	Inhibited PD-1, T-cell activation	2021	dMMR, EnC
PD-L1	Atezolizumab	Inhibited PD-L1, T-cell activation	2016	BC, HCC, melanoma, NSCLC, SCLC, UC
	Avelumab	Inhibited PD-L1, T-cell activation	2017	MCC, RCC, UC
	Durvalumab	Inhibited PD-L1, T-cell activation	2018	NSCLC, SCLC, UC

CRC, colorectal cancer; HCC, hepatocellular carcinoma; NSCLC, non-small cell lung cancer; RCC, renal cell carcinoma; BCC, basal cell carcinoma; CSCC, cutaneous squamous cell carcinoma; EnC, endometrial carcinoma; EsC, esophageal carcinoma; GC, gastric carcinoma; SCC, squamous cell carcinoma; HL, Hodgkin lymphoma; HNSCC, head, and neck squamous cell carcinoma; UC, urothelial carcinoma BC; Breast cancer; CVC, cervical cancer; MCC, Merkel cell carcinoma; LBCL, large B cell lymphoma; SCLC, small cell lung cancer; dMMR, mismatch repair deficient.

Table 1.
Immune checkpoint inhibitors approved by FDA.

and recruiting inhibitory effectors [51]. Moreover, CTLA4 promotes the internalization of its ligands, which prevents them from binding to CD28 and reduces IL-2 production and T-cell proliferation [52].

4.2 Clinical application of CTLA-4 inhibitor

In 2010, the FDA authorized ipilimumab for advanced melanoma treatment, making it the first medicine with a survival advantage for metastatic melanoma. Long-term studies have shown that antitumor immunity persists following CTLA4 inhibition, confirming the stability of this survival effect [53]. Unfortunately, findings from trials in renal cell carcinoma [54], non-small-cell lung cancer [55], small-cell lung cancer [56], and prostate cancer [57] were less effective than those reported in melanoma patients. The FDA has not yet approved tremelimumab, an IgG2 isotype CTLA4-blocking antibody, because it did not extend survival in patients with advanced melanoma. It is believed that the effectiveness of ipilimumab and tremelimumab differs due to differences in binding kinetics and the ability to mediate cytotoxicity [58].

4.3 PD-1/PD-L1

PD1 was initially considered to be a potential modulator of apoptosis. However, later data revealed a role in regulating hyperactivation of the immune system, like CTLA4 [59]. CTLA4 and CD28 have 20% and 15% amino acid identity, respectively [60]. CTLA4 limits T-cell activation in peripheral tissues, whereas PD1 regulates T-cell activation predominantly within lymphoid organs. Relatively, the PD1 axis performs a particular role in self-tolerance in T cells [47].

4.4 Clinical application of PD-1/PD-L1 inhibitor

Pembrolizumab and nivolumab (both IgG4), humanized and completely human anti-PD1 monoclonal antibodies (mAbs), were the first FDA-approved PD1-targeted therapies for refractory and unresectable melanoma in 2014 [61]. Pembrolizumab exceeded ipilimumab in six-month progression-free survival and overall survival [62]. Pembrolizumab was approved in 2015 for treating non-small-cell lung cancer because it increased progression-free survival by 4.3 months compared to platinum-based chemotherapeutics and was more productive than paclitaxel [63]. However, the different organs have distinct immunosuppressive microenvironments, so it's difficult to anticipate which patients may benefit. Nivolumab has since been approved for treating renal cell carcinoma, head and neck squamous cell carcinoma, urothelial carcinoma, hepatocellular carcinoma, Hodgkin lymphoma, and colorectal cancer, similar to pembrolizumab [64].

PDL1 also targeted several antibodies that have benefits for cancer treatment. Atezolizumab (an IgG4 antibody), the first PDL1-targeted humanized mAbs, was licensed to treat urothelial cancer in 2016 [65]. However, more trial results have not shown that atezolizumab has clinical efficacy in urothelial carcinoma over the standard of treatment, even though it is less toxic than conventional chemotherapy [66]. In 2017, avelumab and durvalumab, two new anti-PDL1 human mAbs, were introduced to the market [67]. As a result, similar to PD1, blocking PDL1 is successful in difficult-to-treat cancers.

4.5 Clinical challenges during the blockade of immune checkpoint

Potent immune effector mechanisms are activated by blocking a naturally occurring immunological checkpoint [68]. According to a meta-analysis study, immune-related adverse events are expected to occur in 15–90% of patients. However, patients treated with CTLA4 and PD1 inhibitors have more severe episodes that require intervention in 15–30% of cases [69]. In addition, patients receiving anti-CTLA4 medication are at a higher risk of hypothyroidism, hepatotoxicity, and pneumonitis. In contrast, the patient that receives PD1-targeted drugs is more likely to develop hypothyroidism, hepatotoxicity, and pneumonitis [70].

Adverse events are particularly problematic in adjuvant chemotherapy because late-onset, frequently severe toxicities can impact tumor-survivor patients even after surgery only [71]. However, the toxicity of immune checkpoint inhibitors is more tolerable than that of conventional chemotherapeutics [72].

5. Conclusion and future perspectives

A decade after immune checkpoint protein discovery, such as PD-1/PDL-1 and CTLA-4, still offers hope for the cure for cancer patients. Although not all patients may benefit from these medications, some of the drugs demonstrated dose-dependent adverse events of mild to moderate. Unfortunately, not every cancer type responds well to the treatment, and the only options are to discontinue therapy or switch to conventional cancer treatments.

Even though several studies showing the side effects of ICI do not discourage researchers from developing the new finding, these drugs will probably be examined in adjuvant or neoadjuvant approaches, based on clinical responses in various cancer types, to increase the overall survival of many cancer patients. In addition, understanding the systemic effect mechanism caused by immunotherapy may help gain better knowledge for an effective antitumor response. Finally, combination therapy with various immunotherapy, chemotherapy, targeted medicines, radiation, and T-cell-based therapies can potentially improve the outcomes, especially in patients who have not responded well to immunotherapy-based treatments.

Author details

Poppy Diah Palupi[1*], Mohammed Safwan Ali Khan[2,3] and Nur Dina Amalina[4]

1 Department of Pharmacology and Clinical Pharmacy, Nusaputera College of Pharmaceutical Science, Semarang, Indonesia

2 Faculty of Pharmacy, Department of Pharmacology, Düzce Üniversitesi, Türkiye

3 Faculty of Pharmacy, Department of Toxicology, Cyprus International University, Cyprus

4 Faculty of Mathematics and Natural Sciences, Pharmacy Study Program, Universitas Negeri Semarang, Semarang, Indonesia

*Address all correspondence to: poppydiahp@gmail.com

References

[1] Ferlay J et al. Cancer incidence and mortality worldwide: Sources, methods and major patterns in GLOBOCAN 2012. International Journal of Cancer. 2015;**136**(5):E359-E386. DOI: 10.1002/IJC.29210

[2] Gonzalez H, Hagerling C, Werb Z. Roles of the immune system in cancer: From tumor initiation to metastatic progression. Genes & Development. 2018;**32**(19-20):1267. DOI: 10.1101/GAD.314617.118

[3] Ponomarev AV, Shubina IZ. Insights into mechanisms of tumor and immune system interaction: Association with wound healing. Frontiers in Oncology. 2019;**9**(OCT):1115. DOI: 10.3389/FONC.2019.01115/XML/NLM

[4] NIH. Immunotherapy for Cancer. National Cancer Institute; 2019

[5] Becker JC, Andersen MH, Schrama D, Thor Straten P. Immune-suppressive properties of the tumor microenvironment. Cancer Immunology, Immunotherapy. 2013;**62**(7):1137-1148. DOI: 10.1007/S00262-013-1434-6

[6] Dunn GP, Bruce AT, Ikeda H, Old LJ, Schreiber RD. Cancer immunoediting: From immunosurveillance to tumor escape. Nature Immunology. 2002;**3**(11):991-998. DOI: 10.1038/NI1102-991

[7] Guerrouahen BS, Maccalli C, Cugno C, Rutella S, Akporiaye ET. Reverting immune suppression to enhance cancer immunotherapy. Frontiers in Oncology. 2020;**9**:1554. DOI: 10.3389/FONC.2019.01554/BIBTEX

[8] Fridman WH, Pagès F, Sauts-Fridman C, Galon J. The immune contexture in human tumours: Impact on clinical outcome. Nature Reviews. Cancer. 2012;**12**(4):298-306. DOI: 10.1038/NRC3245

[9] Neophytou CM, Panagi M, Stylianopoulos T, Papageorgis P. The role of tumor microenvironment in cancer metastasis: Molecular mechanisms and therapeutic opportunities. Cancers. 2021;**13**(9):2053. DOI: 10.3390/CANCERS13092053

[10] Vinay DS et al. Immune evasion in cancer: Mechanistic basis and therapeutic strategies. Seminars in Cancer Biology. 2015;**35**(Suppl):S185-S198. DOI: 10.1016/J.SEMCANCER.2015.03.004

[11] Shimizu K, Iyoda T, Okada M, Yamasaki S, Fujii SI. Immune suppression and reversal of the suppressive tumor microenvironment. International Immunology. 2018;**30**(10):445-455. DOI: 10.1093/INTIMM/DXY042

[12] Tanaka A, Sakaguchi S. Regulatory T cells in cancer immunotherapy. Cell Research. 2017;**27**(1):109-118. DOI: 10.1038/CR.2016.151

[13] Sakaguchi S. Naturally arising CD4+ regulatory t cells for immunologic self-tolerance and negative control of immune responses. Annual Review of Immunology. 2004;**22**:531-562. DOI: 10.1146/ANNUREV.IMMUNOL.21.120601.141122

[14] Shimizu J, Yamazaki S, Sakaguchi S. Induction of tumor immunity by removing CD25+CD4+ T cells: A common basis between tumor immunity and autoimmunity. Journal of Immunology. 1999;**163**(10):5211-5218

[15] Sakaguchi S, Yamaguchi T, Nomura T, Ono M. Regulatory T cells and immune tolerance. Cell. 2008;**133**(5):775-787. DOI: 10.1016/J.CELL.2008.05.009

[16] Yamaguchi T et al. Construction of self-recognizing regulatory T cells from conventional T cells by controlling CTLA-4 and IL-2 expression. Proceedings of the National Academy of Sciences of the United States of America. 2013;**110**(23):E2116-E2125. DOI: 10.1073/PNAS.1307185110

[17] Negi S, Saini S, Tandel N, Sahu K, Mishra RPN, Tyagi RK. Translating treg therapy for inflammatory bowel disease in humanized mice. Cell. 2021;**10**(8):1-28. DOI: 10.3390/cells10081847

[18] Tyagi RK et al. HLA-restriction of human Treg cells is not required for therapeutic efficacy of low-dose IL-2 in humanized mice. Frontiers in Immunology. 2021;**12**(February):1-8. DOI: 10.3389/fimmu.2021.630204

[19] Tönshoff B. Immunosuppressants in organ transplantation. Handbook of Experimental Pharmacology. 2020;**261**:441-469. DOI: 10.1007/164_2019_331

[20] Vial T, Descotes J. Immunosuppressive drugs and cancer. Toxicology. 2003;**185**(3):229-240. DOI: 10.1016/S0300-483X(02)00612-1

[21] Taylor AL, Watson CJE, Bradley JA. Immunosuppressive agents in solid organ transplantation: Mechanisms of action and therapeutic efficacy. Critical Reviews in Oncology/Hematology. 2005;**56**(1):23-46. DOI: 10.1016/J.CRITREVONC.2005.03.012

[22] London NJ, Farmery SM, Lodge JPA, Will EJ, Davidson AM. Risk of neoplasia in renal transplant patients. Lancet. 1995;**346**(8972):403-406. DOI: 10.1016/S0140-6736(95)92780-8

[23] Weaver JL. Establishing the carcinogenic risk of immunomodulatory drugs. Toxicologic Pathology. 2012;**40**(2):267-271. DOI: 10.1177/0192623311427711

[24] Durnian JM, Stewart RMK, Tatham R, Batterbury M, Kaye SB. Cyclosporin-a associated malignancy. Clinical Ophthalmology. 2007;**1**(4):421

[25] Carenco C et al. Tacrolimus and the risk of solid cancers after liver transplant: A dose effect relationship. American Journal of Transplantation. 2015;**15**(3):678-686. DOI: 10.1111/AJT.13018

[26] Sprangers B, Nair V, Launay-Vacher V, Riella LV, Jhaveri KD. Risk factors associated with post–kidney transplant malignancies: An article from the cancer-kidney international network. Clinical Kidney Journal. 2018;**11**(3):315-329. DOI: 10.1093/CKJ/SFX122

[27] Zitvogel L, Tesniere A, Kroemer G. Cancer despite immunosurveillance: immunoselection and immunosubversion. Nature Reviews. Immunology. 2006;**6**(10):715-727. DOI: 10.1038/NRI1936

[28] Axelrad JE, Lichtiger S, Yajnik V. Inflammatory bowel disease and cancer: The role of inflammation, immunosuppression, and cancer treatment. World Journal of Gastroenterology. 2016;**22**(20):4794. DOI: 10.3748/WJG.V22.I20.4794

[29] Münz C, Moormann A. Immune escape by Epstein-Barr virus associated malignancies. Seminars in Cancer Biology. 2008;**18**(6):381-387. DOI: 10.1016/J.SEMCANCER.2008.10.002

[30] Danial NN, Korsmeyer SJ. Cell death: Critical control points. Cell. 2004;**116**(2):205-219. DOI: 10.1016/S0092-8674(04)00046-7

[31] Balkwill F. Tumour necrosis factor and cancer. Nature Reviews. Cancer. 2009;**9**(5):361-371. DOI: 10.1038/NRC2628

[32] Abraham RT, Wiederrecht GJ. Immunopharmacology of rapamycin. Annual Review of Immunology. 2003;**14**:483-510. DOI: 10.1146/ANNUREV.IMMUNOL.14.1.483

[33] Keogh A et al. Sirolimus in De novo heart transplant recipients reduces acute rejection and prevents coronary artery disease at 2 years. Circulation. 2004;**110**(17):2694-2700. DOI: 10.1161/01.CIR.0000136812.90177.94

[34] Trotter JF et al. Liver transplantation using sirolimus and minimal corticosteroids (3-day taper). Liver Transplant. 2001;7(4):343-351. DOI: 10.1053/JLTS.2001.23012

[35] Shitrit D et al. Use of sirolimus and low-dose calcineurin inhibitor in lung transplant recipients with renal impairment: Results of a controlled pilot study. Kidney International. 2005;**67**(4):1471-1475. DOI: 10.1111/J.1523-1755.2005.00224.X

[36] Waldner M, Fantus D, Solari M, Thomson AW. New perspectives on mTOR inhibitors (rapamycin, rapalogs and TORKinibs) in transplantation. British Journal of Clinical Pharmacology. 2016;**82**(5):1158. DOI: 10.1111/BCP.12893

[37] Arcaro A, Guerreiro AS. The phosphoinositide 3-kinase pathway in human cancer: Genetic alterations and therapeutic implications. Current Genomics. 2007;**8**(5):271. DOI: 10.2174/138920207782446160

[38] Blagosklonny MV. Hypoxia, MTOR and autophagy: Converging on senescence or quiescence. Autophagy. 2013;**9**(2):260-262. DOI: 10.4161/AUTO.22783

[39] Blagosklonny MV. Rapalogs in cancer prevention: Anti-aging or anticancer? Cancer Biology & Therapy. 2012;**13**(14):1349. DOI: 10.4161/CBT.22859

[40] Grabiner BC et al. A diverse array of cancer-associated mTOR mutations are hyperactivating and can predict rapamycin sensitivity. Cancer Discovery. 2014;**4**(5):554. DOI: 10.1158/2159-8290.CD-13-0929

[41] Lebien TW, Tedder TF. B lymphocytes: How they develop and function. Blood. 2008;**112**(5):1570-1580. DOI: 10.1182/BLOOD-2008-02-078071

[42] He X, Xu C. Immune checkpoint signaling and cancer immunotherapy. Cell Research. 2020;**30**(8):660-669. DOI: 10.1038/s41422-020-0343-4

[43] Waldman AD, Fritz JM, Lenardo MJ. A guide to cancer immunotherapy: From T cell basic science to clinical practice. Nature Reviews. Immunology. 2020;**20**(11):651-668. DOI: 10.1038/s41577-020-0306-5

[44] Pardoll DM. The blockade of immune checkpoints in cancer immunotherapy. Nature Reviews. Cancer. 2012;**12**(4):264. DOI: 10.1038/NRC3239

[45] Whiteside TL. The tumor microenvironment and its role in promoting tumor growth. Oncogene. 2008;**27**(45):5904-5912. DOI: 10.1038/ONC.2008.271

[46] Marin-Acevedo JA, Kimbrough EMO, Lou Y. Next generation of

immune checkpoint inhibitors and beyond. Journal of Hematology & Oncology. 2021;**14**(1):45. DOI: 10.1186/S13045-021-01056-8

[47] Fife BT, Bluestone JA. Control of peripheral T-cell tolerance and autoimmunity via the CTLA-4 and PD-1 pathways. Immunological Reviews. 2008;**224**(1):166-182. DOI: 10.1111/J.1600-065X.2008.00662.X

[48] Rudd CE, Taylor A, Schneider H. CD28 and CTLA-4 coreceptor expression and signal transduction. Immunological Reviews. 2009;**229**(1):12-26. DOI: 10.1111/J.1600-065X.2009.00770.X

[49] Dariavach P, Mattéi M-G, Golstein P, Lefranc M-P. Human Ig superfamily CTLA-4 gene: Chromosomal localization and identity of protein sequence between murine and human CTLA-4 cytoplasmic domains. European Journal of Immunology. 1988;**18**(12):1901-1905. DOI: 10.1002/EJI.1830181206

[50] Pentcheva-Hoang T, Egen JG, Wojnoonski K, Allison JP. B7-1 and B7-2 selectively recruit CTLA-4 and CD28 to the immunological synapse. Immunity. 2004;**21**(3):401-413. DOI: 10.1016/J.IMMUNI.2004.06.017

[51] Intlekofer AM, Thompson CB. At the bench: Preclinical rationale for CTLA-4 and PD-1 blockade as cancer immunotherapy. Journal of Leukocyte Biology. 2013;**94**(1):25-39. DOI: 10.1189/JLB.1212621

[52] Qureshi OS et al. Trans-endocytosis of CD80 and CD86: A molecular basis for the cell-extrinsic function of CTLA-4. Science. 2011;**332**(6029):600-603. DOI: 10.1126/SCIENCE.1202947

[53] Maio M et al. Five-year survival rates for treatment-naive patients with advanced melanoma who received ipilimumab plus dacarbazine in a phase III trial. Journal of Clinical Oncology. 2015;**33**(10):1191-1196. DOI: 10.1200/JCO.2014.56.6018

[54] Yang JC et al. Ipilimumab (anti-CTLA4 antibody) causes regression of metastatic renal cell cancer associated with enteritis and hypophysitis. Journal of Immunotherapy. 2007;**30**(8):825-830. DOI: 10.1097/CJI.0B013E318156E47E

[55] Lynch TJ et al. Ipilimumab in combination with paclitaxel and carboplatin as first-line treatment in stage IIIB/IV non-small-cell lung cancer: Results from a randomized, double-blind, multicenter phase II study. Journal of Clinical Oncology. 2012;**30**(17):2046-2054. DOI: 10.1200/JCO.2011.38.4032

[56] Reck M et al. Ipilimumab in combination with paclitaxel and carboplatin as first-line therapy in extensive-disease-small-cell lung cancer: Results from a randomized, double-blind, multicenter phase 2 trial. Ann. Oncol. Off. J. Eur. Soc. Med. Oncol. 2013;**24**(1):75-83. DOI: 10.1093/ANNONC/MDS213

[57] Kwon ED et al. Ipilimumab versus placebo after radiotherapy in patients with metastatic castration-resistant prostate cancer that had progressed after docetaxel chemotherapy (CA184-043): A multicentre, randomised, double-blind, phase 3 trial. The Lancet Oncology. 2014;**15**(7):700-712. DOI: 10.1016/S1470-2045(14)70189-5

[58] He M et al. Remarkably similar CTLA-4 binding properties of therapeutic ipilimumab and tremelimumab antibodies. Oncotarget. 2017;**8**(40):67129-67139. DOI: 10.18632/ONCOTARGET.18004

[59] Ishida Y, Agata Y, Shibahara K, Honjo T. Induced expression of PD-1, a novel member of the immunoglobulin gene superfamily, upon programmed cell death. The EMBO Journal. 1992;**11**(11):3887-3895. DOI: 10.1002/J.1460-2075.1992.TB05481.X

[60] Carreno BM, Collins M. The B7 family of ligands and its receptors: New pathways for costimulation and inhibition of immune responses. Annual Review of Immunology. 2002;**20**:29-53. DOI: 10.1146/ANNUREV.IMMUNOL.20.091101.091806

[61] Gong J, Chehrazi-Raffle A, Reddi S, Salgia R. Development of PD-1 and PD-L1 inhibitors as a form of cancer immunotherapy: A comprehensive review of registration trials and future considerations. Journal for Immunotherapy of Cancer. 2018;**6**(1):8. DOI: 10.1186/S40425-018-0316-Z

[62] Robert C et al. Pembrolizumab versus Ipilimumab in Advanced Melanoma. The New England Journal of Medicine. 2015;**372**(26):2521-2532. DOI: 10.1056/NEJMOA1503093

[63] Herbst RS et al. Pembrolizumab versus docetaxel for previously treated, PD-L1-positive, advanced non-small-cell lung cancer (KEYNOTE-010): A randomised controlled trial. Lancet (London, England). 2016;**387**(10027):1540-1550. DOI: 10.1016/S0140-6736(15)01281-7

[64] Overman MJ et al. Nivolumab in patients with metastatic DNA mismatch repair-deficient or microsatellite instability-high colorectal cancer (CheckMate 142): An open-label, multicentre, phase 2 study. The Lancet Oncology. 2017;**18**(9):1182-1191. DOI: 10.1016/S1470-2045(17)30422-9

[65] Rosenberg JE et al. Atezolizumab in patients with locally advanced and metastatic urothelial carcinoma who have progressed following treatment with platinum-based chemotherapy: A single-arm, multicentre, phase 2 trial. Lancet. 2016;**387**(10031):1909-1920. DOI: 10.1016/S0140-6736(16)00561-4

[66] Powles T et al. Atezolizumab versus chemotherapy in patients with platinum-treated locally advanced or metastatic urothelial carcinoma (IMvigor211): A multicentre, open-label, phase 3 randomised controlled trial. Lancet. 2018;**391**(10122):748-757. DOI: 10.1016/S0140-6736(17)33297-X

[67] Hargadon KM, Johnson CE, Williams CJ. Immune checkpoint blockade therapy for cancer: An overview of FDA-approved immune checkpoint inhibitors. International Immunopharmacology. 2018;**62**:29-39. DOI: 10.1016/J.INTIMP.2018.06.001

[68] Fritz JM, Lenardo MJ. Development of immune checkpoint therapy for cancer. The Journal of Experimental Medicine. 2019;**216**(6):1244-1254. DOI: 10.1084/JEM.20182395

[69] Michot JM et al. Immune-related adverse events with immune checkpoint blockade: A comprehensive review. European Journal of Cancer. 2016;**54**:139-148. DOI: 10.1016/J.EJCA.2015.11.016

[70] Kumar V, Chaudhary N, Garg M, Floudas CS, Soni P, Chandra AB. Current diagnosis and Management of Immune Related Adverse Events (irAEs) induced by immune checkpoint inhibitor therapy. Frontiers in Pharmacology. 2017;**8**(FEB):49. DOI: 10.3389/FPHAR.2017.00049

[71] de Miguel M, Calvo E. Clinical challenges of immune checkpoint inhibitors. Cancer Cell.

2020;**38**(3):326-333. DOI: 10.1016/J.CCELL.2020.07.004

[72] Nishijima TF, Shachar SS, Nyrop KA, Muss HB. Safety and tolerability of PD-1/PD-L1 inhibitors compared with chemotherapy in patients with advanced cancer: A meta-analysis. The Oncologist. 2017;**22**(4):470-479. DOI: 10.1634/THEONCOLOGIST.2016-0419

Section 3

Infectious, Inflammatory Diseases and Immunomodulation

Chapter 7

Perspective Chapter: Role of Cytotrophoblast Cells and Placenta-Derived Exosomes in Regulatory B Cell Differentiation and Function during Pregnancy

Gatien A.G. Lokossou and Maximin Senou

Abstract

Pregnancy is a particular physiologic stage during which immune regulation is essential. A successful placentation and subsequent fetal development depend on the delicate balance between moderate pro-inflammatory response and immune tolerance. Findings have pointed out a crucial role for regulatory B cells (Bregs) in establishing an immunomodulatory (IM) environment relevant to pregnancy. In a steady state, Bregs represent 10% of B cells in peripheral blood, a proportion that increases during pregnancy, with the highest rate being observed in *post-partum*. In the context of pregnancy, Bregs seem to be well positioned to perform the mechanisms that accommodate the growing semi-allogenic fetus and also allow the adequate immune response to pathogen. This chapter discusses the mechanism of action of Bregs during human pregnancy. Also, we will evoke interactions between maternal immune cells and fetal annexes that result in hijacking the naïve B cells to educate and to differentiate them into Bregs.

Keywords: pregnancy, immunoregulation, exosomes, cytotrophoblast, Bregs

1. Introduction

Preeclampsia (PE) is a placental disorder affecting 2–8% of pregnancies with the highest burden observed in poor countries [1–4]. PE and severe PE are characterized by exacerbated pro-inflammatory (PI) responses, leading to significant maternal and perinatal morbidity and mortality [5–8]. Successful placentation and the subsequent fetal development depend on the delicate balance between moderate PI response and immune tolerance.

Recent data has shown that the B cell profile is changed during pregnancy to accommodate the growing fetus [9, 10]. Findings have pointed out a crucial role for regulatory B cells (Bregs) in establishing an immunomodulatory (IM) environment relevant for pregnancy [11]. In a steady state, Bregs represent 10% of B cells in peripheral blood, a proportion which increases during pregnancy, with the highest rate being observed in post-partum [12, 13]. Studies have demonstrated an IM function for IL-10-producing Bregs against ongoing inflammatory events to both limits the infection [14] and promote a successful outcome of pregnancy [15, 16]. It was also shown that early transfer of Bregs from normal pregnant to abortion-prone mice prevented fetal rejection and restored pregnancy tolerance [17].

Indeed, the maternal immune system needs to recognize and accommodate a developing semi-allogeneic fetus. Regulatory T cells are shown to contribute to normal pregnancy, and considering the immune regulatory involvement of Bregs cells in the fields of autoimmunity, transplantation tolerance, and cancer biology, the mechanism underlying Bregs activities during pregnancy needs to be unraveled [18–20]. The immune regulatory function of Bregs consist of inhibition of the differentiation of effector T cells and dendritic cells (DCs), and activation of Tregs [21, 22].

Today, the consensus is not fully established on the characterization of Bregs with respect to cell surface markers. Recent studies have shown that the regulative roles of Bregs are due to the production of the antiinflammatory cytokine interleukin-10 (IL-10) [23–25]. However, recent data indicated that some B cell subsets perform regulatory functions without IL-10 involvement suggesting that other Bregs use multimechanistic to regulate immune responses.

In mice, multiple B cell subsets are identified to play regulatory function and include the marginal-zone B cells, the transitional 2 marginal-zone precursor cells, follicular B cells, $CD5^+CD178^+$ killer B cells, plasma cells, plasmablasts, $CD5^+CD1d^{hi}IL\text{-}10^+$ B cells, $CD5^+$B-1a cells, GIFT-15 B cells, TIM-1^+ B cells, and PD-L1hi B cells [26, 27]. The IL-10-producing Bregs, also called B10 cells have the CD1dhiCD5$^+$ phenotype [28].

In humans, immature B cells, IL-10^+ B cells (B10), GrB$^+$ B cells, Br1 cells, and plasmablasts are identified to have immunosuppressive functions [26]. Previous data in humans have described Bregs as $CD19^+CD24^{hi}CD27^+$ [29], which are analogous to the mouse B10 cells [26] and $CD19^+CD24^{hi}CD38^{hi}$ cells [26] with the ability to produce IL-10 and to express CD80 and CD86 costimulation molecules [30].

This disparity of these regulatory cells suggests that Bregs are not derived from one specific lineage; rather they may become Bregs following exposure to environmental stimuli such as placenta derived-exosomes.

As PE is PI disease and syncytiotrophoblast (STB)-derived exosomes (SDE) contribute to materno-foetal immuno-tolerance, it will be relevant to understand how STB cells and SDE contribute to PE by altering Bregs differentiation and function during human pregnancy. We will discuss whether a disrupted balance of Bregs could increase susceptibility to PE.

2. Cytotrophoblast cells and placenta-derived exosomes in successful placentation and fetal development

The successful pregnancy requires the suitable development of embryo and adequate placentation [31–33]. The appropriate placentation is based on the proper capacity of villous cytotrophoblasts to fuse and form the syncytiotrophoblast which contributes to development of placenta. Therefore, abnormal cytotrophoblast differentiation results in placental-related pregnancy diseases [34–36].

Many cytotrophoblast cell subtypes (cytotrophoblasts (CTBs), extravillous cytotrophoblasts (EVTs) and syncytiotrophoblast (STB)) with different structures and functions are involved in placentation [37, 38]. After implantation of the zygote, trophoblast cells develop from the outer cells which form the wall of the blastocyst, and differentiate into either villous or extravillous trophoblast cells [38]. The STBs are the outer lining of the placenta; fulfill a vast range of role including gas and nutrient exchange between mother and fetus. The trophoblast cell subtypes in addition to secreting hormones and proteins, physically protect the fetus from pathogens [39, 40]. EVTs are invasive trophoblast cells that are important for implantation of the placenta and the development of the fetus [41–43]. Many pregnancy diseases such as PE and intrauterine growth retardation (IUGR) are the consequences of defective placentation [34–36]. Therefore, the normal development of the placenta is based on complex mechanisms of proliferation and differentiation of trophoblast cells [44, 45]. Many factors (e.g., interferon-induced transmembrane protein 1 (IFITM) and Storkhead box 1 (STOX1) SNPs, Syncytins, and factors released by placenta soluble fms-like tyrosine kinase-1 (sFlt-1), placenta growth factor (PlGF), transforming growth factor-β (TGF-β)) govern the regulation of cytotrophoblast cell differentiation showing their potential use as biomarker [46–48].

Placental syncytialization is maintained throughout pregnancy by the fusion of adjacent CTBs [49, 50] and is important for successful pregnancy [49, 51–54]. Syncytins are important players during syncytialization and *Vargas et al.* and Lokossou et al. indicated that insufficient Syncytin-2 (Syn-2) expression could be the potential cause of PE, shedding light on the correlation between Syn-2 level and cytotrophoblasts fusion [46, 50, 55, 56].

In recent years, the role of exosomes in the development of the placenta has become more and more precise [46, 56, 57]. These microvesicles with diameters of 20–130 nm are extracellular secreted vesicles and are involved in cell-to-cell communication [58]. They, therefore, affect cytotrophoblast differentiation and immune regulation, especially during pregnancy [58, 59]. Exosomes are secreted by most cells and embedded in various substances including proteins [46, 56, 60, 61], mRNA and miRNA [62], and DNA [63]. They can be transported to distant organs and are thought to modify various cells and organ functions [56, 64, 65]. SDE which embedded placenta-specific molecules, including Syn-2, were involved in CTB fusion [55], embryo implantation *via* the promotion of T regulatory cells, suppression of Nuclear Factor-kB signaling pathway [66] and thereby in immune reaction and inflammatory response [56]. Secreted exosomes from the placenta into the systemic circulation lead in multisystemic organ damage, in patients with PE [67]. Reduction of Syn-2 levels in exosomes is suggested to be an early biomarker of PE [46, 60]. Indeed, the identification of women at high risk of PE before its onset is especially a challenge. Exosomes miRNA pattern also appears to be used for early PE diagnosis [68]. SDE in preeclamptic placentas are thought to embed high concentrations of PE-specific contents, resulting in unfavorable microenvironments for the invasion of EVTs and the remodeling of spiral arteries for adequate placentation [46, 57, 63, 66–72].

Nowadays, evidence suggests that disruption of placentation characterizes the pathogenesis of PE [55, 73]. Indeed, STB-derived exosomes are found in maternal circulation [72], and affected endothelial function due to their abundant sFlt-1 and soluble endoglin (sEng) content [69]. These vesicles are also endowed with immune regulation capacities during pregnancy due to Syn-2 embedded in exosomes [56].

Placental exosomes are therefore able to deliver many molecules including proteins around CTB, inducing a particular environment that affects placenta and fetal growth.

The immunosuppressive protein derived from human endogenous retrovirus sequences, Sync-2. plays a leading role in placenta formation [49, 50]. For several years our knowledge has grown on PE and placental exosomes. We have demonstrated the role of Sync-2 in placentation and in T cell immunosuppression [56, 60, 74] during normal pregnancy and PE, suggesting that Sync-2 could be used as an early biomarker of PE. Our recent data from Benin show a gradual diminution, between 7 and 10 weeks of pregnancy (WP), in the incorporation of Sync-2 in serum-derived exosomes from women who had developed a PE later during their pregnancy in comparison to samples from women with normal pregnancy [46]. Sync-2 through its immunosuppressive domain might contribute greatly to creating an immunosuppressive environment. This environment is reinforced and maintained by other factors such as Bregs. This immunosuppressive environment is essential at the beginning of pregnancy for the allograft tolerance constituted by the fetus [75]. As we demonstrated that Sync-2 generates an immunosuppression (IS) environment, Bregs should be important to maintain the IS environment and to prevent allograft rejection. Indeed, PE is a placental and inflammatory disease and syncytiotrophoblast-derived exosomes contribute to materno-fetal immuno-tolerance [56]. Such defective placentation is thought to be caused by an abnormal CTB fusion due to defective production of Sync-2 [46, 49, 55] but also to abnormal maternal immune regulation, involving Sync-2 [56]. Many immune cells, including T cell, macrophage, natural killer, regulatory B and T cells, are also affected during PE [76]. Therefore, it will be of great importance to understand how cytotrophoblast and/or syncytiotrophoblast cells and placenta-derived exosomes contribute to PE by altering Bregs differentiation and function during human pregnancy. By demonstrating that Bregs frequency and function increase susceptibility to PE, would lead to the immediate management of pregnant women predisposed to the development of severe PE and reduce the number of resulting morbidity and deaths.

3. Preeclampsia: state of knowledge

Preeclampsia (PE) is the most common placental disorder affecting pregnancy [77]. PE is associated with vascular dysfunction and deregulated inflammation, oxidative stress, and endothelial dysfunctions [78–80]. This chronic inflammation begins early in pregnancy as a result of stimulation of maternal immune response by trophoblasts and trophoblasts derived-products. It is associated with leukocyte activation, vascular activation and dysfunction and high serum levels of cytokines such as Tumor Necrosis Factor-α (TNF-α) [81–83]. Changes in the levels of immune factors (e.g., cytokines, chemokines) are followed by changes in blood coagulation factors, and apoptotic markers [78, 84–87]. Leukocyte activation is driven by TNF-α, and monocyte-derived cytotoxic protein, that also induces vascular endothelial adhesion molecules. Indeed, increased TNF-α levels in early pregnancy can increase the expression of intercellular adhesion molecule-1 (ICAM-1) on vascular endothelial cells (ECs) and trophoblasts, thereby activating them. Consequently, coagulation cascade, vascular tone, and permeability are disturbed. Moreover, chronic inflammation also activates lymphocyte function-associated antigen-1 (LFA-1) on leukocytes, resulting in the above consequences [84, 85, 88, 89]. This chronic inflammation is also associated with oxidative stress [90] which increases the adhesion of leukocytes to the vascular endothelium and the release of cytokines and anti-angiogenic molecules. Adhesion molecules (e.g., soluble E-selectin and soluble ICAM-1) and reactive oxygen species level are increased in blood collected early in pregnancy from women who develop later PE [83, 91]. During PE, abnormal levels of anti-vascular growth factors (e.g., sFlt-1 and sEng)) lead to maternal vascular inflammatory syndrome characteristics [58]. These antiangiogenic factors induce the decrease of angiogenic placental growth factor (PlGF), poorly affecting angiogenesis during placentation [92]. Defective placenta secreted PlGF into the maternal circulation as a result of impaired endovascular invasion by trophoblast cells and is underlying by cellular oxidative or endoplasmic reticulum stress [93]. Therefore, the level of these factors in the maternal peripheral blood might enable an early diagnosis of PE. Nevertheless, these methods are questioned and did not allow an early prediction (i.e., during the first trimester) of the occurrence of PE [94, 95].

Commonly, PE results in multi-organ failure (e.g., renal insufficiency, liver dysfunction, neurological or hematological complications, uteroplacental dysfunction) in the mother and poor perinatal outcome. Thereby, PE results in significant maternal and perinatal morbidity and mortality [77]. PE affects 2 to 8% of pregnancies and low and middle-income countries (LMIC) are mostly affected [1–3]. PE is defined as new onset hypertension arising after 20 weeks' gestation, but can also occur at a later stage, i.e., 4–12 weeks postpartum [96, 97].

PE is a consequence of failure of paternal antigen-specific tolerance. Moreover, first pregnancy, first pregnancy after partner change, and long interval between pregnancies increase PE risk [98, 99]. The reduced opportunities for exposure to seminal plasma, and pregnancy by sperm or oocyte donation or pregnancies with donated greatly increase PE risk [100–103].

4. Immune regulation and immune tolerance during pregnancy

Contrary to what would be expected, the maternal immune system does not reject the semi-allogeneic fetus allowing the maintenance of the pregnancy [104], but still reacting with infectious agents. This is the result of interplay between maternal and fetal cells creating a tolerogenic microenvironment at the feto-maternal interface [105]. During PE, immunotolerance to fetal antigens, (e.g. trophoblast) is impaired, resulting in disrupted remodeling of the spiral artery and thereby poor placentation. Regulatory T (Tregs) cells play a crucial role in this tolerogenic microenvironment [106, 107] and the maintenance of pregnancy depends on the balance between Tregs and cytotoxic T cells. The ability of fetal cells to escape destruction by maternal immune cells is based on the expression of human leukocyte antigen (HLA)-C molecules by EVTs alone. Therefore, maternal T-cell reactivity to fetal cells is reduced [18]. The T cells immunosuppression during pregnancy is also mediated by HLA-G, E, F and programmed cell death ligand 1 (PD-L1), indoleamine 2,3-dioxygenase (IDO) expression by EVTs [18–20, 108–110]. EVTs also induce T cell suppression by expressing cytokines including IL-10, TGF-β, and IL-35 [12, 13]. At the beginning of pregnancy, $CD56^{bright}CD16^{-}$ decidual natural killer (NK) cells (dNK cells) represent more than 60% of decidual immune cell and express high level of immunosuppressive receptors [111, 112]. Immunotolerance is maintained at the feto-maternal interface by interaction between HLA-G expressed on EVTs and dNK cells and dNK derived-cytokines [57, 113, 114]. To maintain immunotolerance at the feto-maternal interface, between dendritic cells (DCs) and Tregs [115], the cross-presentation of paternal antigens to maternal cytotoxicity T CTLs and $CD4^{+}$ T cells [115, 116] is altered by fetal antigen-specific Tregs at the feto-maternal interface [117–121]. Seminal plasma components such as TGF-β, prostaglandins, MHCs, and minor antigens also functionally affect maternal antigen-presenting cells (APC). These functional changes were maintained by EVTs [103, 122]. In mice, these functional changes favor fetal-antigen-specific Tregs cell expansion in the uterus and uterine drainage lymph nodes [123–125]. Moreover, in mice, PD-L2-expressing dendritic cells (DCs) increase during implantation in allogeneic pregnancy in mice [126]. These cells limit inflammation whereas, the decrease of M2 macrophages, which inhibit inflammation and promote tissue repair, results in implantation failure in mice [127]. Furthermore, *in vitro*, the close interaction between decidual macrophages and EVTs results in an increased number of Tregs after co-culture with peripheral blood-derived $CD4^+$ T cells [128, 129]. These results show clearly that EVTs oriented the differentiation of $CD4^+$ conventional T cells into antigen-specific peripheral Tregs [121].

Miscarriages, PE, and implantation failure are some characteristics of pregnancy complications. The dysfunction of Tregs is clearly involved [121, 130]. Indeed, in recurrent pregnancy loss and during PE, low level of Tregs in the peripheral blood and uterus has been reported [121]. A recent study has shown that during PE, clonally expanded effector Tregs were significantly decreased in the decidua compared with normal pregnancy, suggesting an insufficient Tregs antigen-specific tolerance [131].

In addition to this induced-immunosuppression, decidual CTLs were also suppressed by EVTs and other immune cells; to allow a good course of pregnancy, without suppression of CTLs functions against virus [117]. During late gestation, le level of PD-L1 on clonally expanded CTLs increases significantly compared to the beginning of pregnancy, showing that strong suppressive signals are necessary to inhibit the allo-reaction by CTLs in the late gestational period [119]. Moreover, during late onset of PE, the level of PD-L1 on clonal CTLs decreased compared with that in normal pregnancy [119], suggesting insufficient suppression of antigen-specific CTLs in PE.

Overall, during normal pregnancy although Tregs and CTLs recognize fetal antigens at the feto-maternal interface, antigen-specific Tregs induce tolerance, while the cytotoxic function of CTLs is suppressed. Therefore, the imbalance of suppressive role of Tregs and activation of CTLs is likely associated with PE.

In PE, type-1 T helper (Th1) cells numbers are also increased [73] and secrete pro-inflammatory cytokines, such as TNF-α, interferon-γ (IFN-γ), and interleukin (IL)-6. Increased Th17 cells secreting the pro-inflammatory cytokine IL-17 are also found during PE [83, 132].

5. Regulatory B cells establish immunomodulatory environment for pregnancy

With regard to their multi-faceted roles, B cells may participate in successful pregnancy [11]. A subtype of B cells named Bregs exhibits immunosuppressive function and therefore is considered an important players in immunological tolerance during pregnancy. During pregnancy, the change of the maternal immune response is governed by a range of cytokines that shape the type and abundance of leukocyte subsets in the decidua and placenta. In addition, these changes also include the reduction of antigen-presenting capacities of monocytes, macrophages, and DCs; inhibition of NK cells, T cells, and B cells; proliferation of dNK cells; maintenance of tolerogenic DCs; and the induction of Tregs [133].

Regulatory roles of Bregs were attributed exclusively to the production of the anti-inflammatory cytokine interleukin-10 [23, 29, 134], even recent data have identified other B cell subsets with regulatory functions without IL-10 production. Indeed,

Bregs are cells that dampen ongoing inflammatory events in murine models [15] and counteract excessive pro-inflammatory responses during infection [14]. IL-10 is crucial for optimal pregnancy outcomes, Therefore IL-10 deficiency is related to fetal resorption, growth restriction, and even death of mother and child [135, 136]. This antiinflammatory cytokine is found in high levels in the decidual and placenta during pregnancy and is involved in damping the pro-inflammatory cytokine response. Interesting fact, at the beginning of pregnancy, the inflammation induced by the recognition of paternal antigens is upset by the production of IL-10 [137].

Bregs proportion increases during pregnancy and *in postpartum* [138] and first-trimester peripheral blood-derived Bregs are shown to inhibit TNF-α secretion by activated T effector cells [139]. In the context of pregnancy, Bregs seem to be well positioned to perform the mechanisms that accommodate the growing semi-allogeneic fetus and also allow the adequate immune response to the pathogen [10]. However, the mechanism of action of Bregs during pregnancy remains curtailed even if their importance in pregnancy has been shown in mouse models [138].

In order to allow a successful placentation and suitable fetal development, the maternal immune system undergoes several changes while allowing the mother to defend herself against infections [10]. A German group has developed a PE mouse model by transferring activated Th1-like splenocytes into normal pregnant mice and has demonstrated that pregnancy-associated immuno-regulation involved a shift from inflammatory toward anti-inflammatory immune responses mainly controlled by T and B cells [140, 141]. They have reported that Bregs were active players in the maintenance of pregnancy by modulating T cell functions [142]. Indeed, they have shown that Bregs transfer from normal pregnant to abortion-prone mice early in pregnancy prevents fetal rejection and restores pregnancy tolerance in mice [17].

The action of Bregs during pregnancy is interconnected with that of Tregs and DCs, providing an appropriate environment for fetal growth.

As described above, both Tregs and DCs play critical roles in determining pregnancy outcomes. In normal pregnancy, fetal-tolerant involves decidual DCs that remain immature and knew as tolerogenic DCs [143]. Tregs are also crucial players in maintaining maternal-fetal immune tolerance. At the onset of embryo implantation, expansion of the Tregs population improves the outcome of pregnancy whereas deficiency or low numbers of Tregs in the uterus during pregnancy leads to pregnancy lost (i.e., abortion or miscarriage) [22, 144].

As anti-inflammatory signal such as IL-10 or TGF-β is crucial to maintain the DC immature phenotype that prevents the activation of T cells [145], IL-10 and or TGF-β must then be initially: 1- present in the uterus and placenta to either drive the induction of immune tolerant DCs followed by the induction of Tregs, or 2- directly drive the induction of Tregs.

Overall Breg's role in pregnancy seems to be upstream to that of Tregs [17] and therefore, Bregs seem to be the first sources of IL-10 and TGF-β [10, 17].

6. Impact of cytotrophoblast cells and placenta-derived exosomes on regulatory B cells differentiation and function

The pathophysiology of PE is poorly understood despite the evidences supporting a role of immune system in the development of PE [56]. B cells represent a dominant component in the pathogenesis of PE and studies focused on the number and functions of Bregs during PE are of great interest in understanding the pathophysiology of PE [139]. Harnessing Bregs functions may lead to the capacity of using Bregs as an immunotherapeutic agent for averting and treating pregnancy pathologies such as PE. Perinatal cells including cells from term placenta and fetal annexes (amniotic membrane, chorionic membrane, umbilical cord, etc) are able to inhibit B cell proliferation, impair B cell differentiation and promote Bregs formation, frequently due to bioactive factors secreted by perinatal cells [11, 146]. These cells are considered as a promising tool for therapeutic approaches in PE [146]. Interactions between maternal immune cells and fetal annexes may result in hijacking naïve B cells and educating them to become Bregs. However, how cytotrophoblast (CT) and/or syncytiotrophoblast (ST) cells regulate Bregs differentiation and function during pregnancy is still unknown. Maybe in case of PE, CT and ST and their derived-vesicles (e.g. exosomes) will prevent adequate Bregs development and function, resulting in reduced and dysfunctional Bregs. This default of Bregs might result in an inflammatory environment, which will increase the susceptibility to PE.

Recent *in vitro* and *in vivo studies* have shown that perinatal cells and perinatal cells derived-vesicles interfere with the activation and differentiation of innate and adaptive immune system cells [11]. Poor knowledge is available about the impact of perinatal cells on B lymphocytes, even if some of the complex cross-talks between perinatal cells and B cells have been described. These studies demonstrated that perinatal cells have a strong antiproliferative capacity on B cells, but were not based on cell–cell contact. The demonstration is based on bioactive factors secreted by perinatal cells. For instance, co-cultured human mesenchymal stromal cells (MSC) isolated from umbilical cord (hUC-MSC) in a contact independent with mouse splenic B cells result in abrogation of the proliferation of activated B cells [147]. Likewise, human umbilical cord matrix cells co-cultured with a B cell cancer line (i.e. Burkitt's lymphoma cell line) [148], or with auto-reactive B cells from PBMC of immune thrombocytopenic patients results in inhibition of these B cells proliferation [149]. These observations were confirmed by using other perinatal cells (e.g., mesenchymal stromal cells (MSC)) purified from the amniotic membrane (hAMSC). This MSC supernatant is able to suppress $CD19^+$ B cell proliferation in PBMC or purified B cells from PBMC, confirming that cell-to-cell contact was not required and suggesting the role of soluble molecules and vesicles such as exosomes [11]. Similarly, human amniotic fluid stromal cells and their conditioned medium (CM) strongly suppress B cell activation and proliferation, and significantly inhibited the expression of CD80/CD86 costimulatory molecules on activated B lymphocytes [150].

However, some data contradict these observations and showed that human amniotic fluid stromal cells are able to suppress the apoptosis of B lymphocytes, favoring an increase in activated B cell survival. The mechanism underlying this inhibition is based on the decrease of the expression of the negative co-inhibitory molecules B7 homolog 4 (B7H4) and programmed death-ligand 1 (PD-L1) on activated B lymphocytes [150]. Moreover, an increase in B cell proliferation and a reduction in spontaneous apoptosis in the presence of human amniotic epithelial cells (hAEC) were also described [143]. Umbilical cord derived-MSC were not able to affect [144] or in other studies able to highly induce the *in vitro* growth of PBMC derived-B cells [145].

It is also demonstrated that human amniotic fluid stromal cells induce down-regulation of the proportion of B1 cells [150], resulting in the reduction of the B cell subset mainly involved in the production of autoantibodies in PE [151–153]. Many studies have shown that perinatal cell and their CM are able to block antibody-secreting cells $CD19^+CD27^+CD38^+$ and the differentiation of B cells into $CD138^+$ plasma cells, resulting in the reduction of secreted immunoglobulin [11, 147, 150]. However, co-culture of purified B cells with human amniotic fluid stromal cells results in reduction of the proportion of $CD19^+CD20^+CD27^+$ memory B cells [150], whereas PBMC cultured in the presence of CM-hAMSC increases $CD19^+CD27^+CD38^-$ memory B cells [11]. These different results may be explained by the presence of other immune cells among PBMC instead purified B cells. Moreover, different conditions of stimulation were used to activate B cells, and the lack of consensus in the markers used to characterize the B cell population could also support the distinct results observed by different groups.

Perinatal cells not only modulate B cell function by favoring their differentiation toward plasma cells, but they also promote the formation of Bregs. Indeed, it was reported that hAEC induced the expansion of $CD19^+CD24^{hi}CD38^{hi}$ Bregs [143]. However, recent data suggested that IL-10^+ Bregs were inhibited by human amniotic fluid stromal cells [150]. These observations clearly showed that more knowledge is needed to understand the impact of perinatal cells and other related vesicles on Bregs differentiation and functions. Thereby, it's important to identify the signaling pathways involved in underlying how perinatal cells and derivatives affect B cell proliferation and differentiation. Two signaling pathways were identified to be suppressed through CpG oligodeoxynucleotides (CpG ODN) by hAMSC: 1-the Toll-like receptor 9 (TLR9)-myeloid differentiation primary response 88 (MyD88)-interleukin-1 receptor-associated kinase (IRAK)1/4 and 2- the TLR9-phosphatidylinositol 3-kinase (PI3K)-protein kinase B (AKT) pathways [11]. This suppression results in a reduction of uptake of the CpG ODN by CD205, TLR9, and CD14. Consequently, IRAK-4, mitogen-activated protein kinases (MAPK) (c-Jun N-terminal Kinase (JNK), p38 MAPK, extracellular signal-regulated kinase (ERK)) and nuclear factor kappa-light-chain-enhancer of activated B cells (NF-kB) pathways were inhibited. This induces an important reduction in the expression of phosphorylated AKT [11, 147]. The exact mechanism by which perinatal cells and derivatives induce Bregs differentiation is still unknown, and needs to be investigated [154]. Few data demonstrate that perinatal cells produce soluble factors including prostanoids (i.e., prostaglandin E2 (PGE2)), and maybe exosomes to immune regulate cells [155–157]. Therefore, we can speculate that Bregs differentiation is also induced by bioactive vesicles.

Based on *in vitro* results showing that perinatal cells have immunomodulatory properties, they were successfully tested in several inflammatory and immune-mediated diseases, including lung [158, 159] and liver [160] fibrosis, inflammatory bowel disease, collagen-induced arthritis, experimental autoimmune encephalomyelitis [161], multiple sclerosis, wound healing [157, 162], traumatic brain injury [163], cerebral ischemia [164], Huntington's disease [165], and diabetes [166].

The therapeutic using hAMSC in pathological conditions driven by B cells has demonstrated a reduced idiopathic pulmonary fibrosis progression [158]. This treatment allows low levels of B cells in alveolar spaces and reduced the amount of $CD138^+$ antibody-secreting cells in lung tissues, suggesting a decrease in B cell recruitment and an impairment of the maturation of B cells. Therapy using hAEC has also shown remarkable results in animal models of Hashimoto's thyroiditis and systemic lupus erythematosus (SLE) [167].

hAEC induced significant up-regulation of Bregs in experimental autoimmune thyroiditis mice. In this experiment, authors have shown that B10 cells are the major target of hAEC. In SLE mice, hAEC has shown the reduction in autoantibody production but without effect on B10 cells, suggesting that the mechanism of hAEC immunomodulation depends on the disease [167].

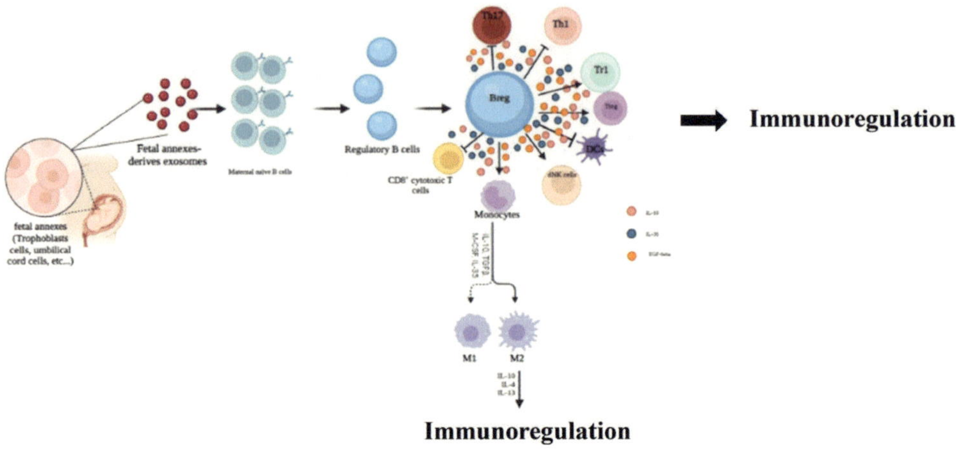

Figure 1.
Differentiation and functional properties of Bregs. Through the production of exosomes by fetal annexes including cytotrophoblast cells, naive B cells can differentiate into Bregs. By producing IL-10, TGF-b, and IL-35, Bregs can suppress tumor necrosis factor-a (TNF-a)-producing monocytes, IL-12-producing dendritic cells, Th17 cells, Th1 cells, and cytotoxic CD8+ T cells. Bregs can also induce the differentiation of immunosuppressive Tregs, T regulatory 1 (Tr1), and dNK cells. This figure was created using Biorender.com.

In the context of chronic graft-versus-host disease (cGVHD) prophylaxis repeated infusion of hUC-MSC seems to minimize the severity and the symptoms of cGVHD by increasing $CD27^{+}$ memory B cells [168].

The immune modulation properties of perinatal cells depend on the origin of these cells. Indeed, fetal-derived cells induce strong inhibition of T-cell proliferation, cytotoxicity, and switch to M2 macrophages, while maternal-derived cells were more strongly able to induce Tregs [169]. **Figure 1** describes the probable implication of exosomes in Bregs differentiation and function (**Figure 1**).

As PE is pro-inflammatory disease and syncytiotrophoblast-derived exosomes (SDE) contribute to materno-fetal immuno-tolerance, it will be useful to understand how STB cells and SDE contribute to PE by altering Bregs differentiation and function during human pregnancy. These mechanisms may be close to those that inhibit immune flares or chronic inflammation in autoimmune diseases and transplantation.

7. Conclusion

Gestation is a remarkable biological process in which the mother carries a fetus harboring half of a foreign genome belonging to the father. To allow the growth of the fetus, the maternal immune system needs to accommodate the semi-allogeneic fetus by dampening its immune responses. This results in a state of immunological tolerance throughout gravidity while maintaining the capacity to respond to pathogens properly. This paradoxical situation requires a perfect regulation of the balance between immune tolerance and immune activation.

To enable more accurate prediction and prevention of PE, its pathogenesis needs to be more understood. Increasing evidence suggests a consequence of the altered immune system in the development of PE. Today, it is clear that perinatal cells have capacity to regulate B cell response at different levels: by inhibiting B cell multiplication, impairing B cell differentiation, and inducing B regulatory cell formation.

Future research should focus on understanding how cytotrophoblast cells and placenta-derived exosomes act on B cells.

Overall, it is clear that cytotrophoblast cells and placenta-derived exosomes harbor the capacity of being a novel therapeutic approach for PE. However, the opposite results and the mainly small number of studies exploring the effect of cytotrophoblast cells and placenta-derived exosomes on the Bregs subset cannot allow deciding on a position. Further *in vitro* and *in vivo* studies are necessary to better decide the immunomodulatory potential of perinatal cells, leading to an important strategy for the treatment of PE.

Conflict of interest

The authors declare no conflict of interest.

Author details

Gatien A.G. Lokossou[1]* and Maximin Senou[2]

1 Research Unit in Applied Microbiology and Pharmacology of Natural Substances, Polytechnic School of Abomey-Calavi, Department Human Biology Engineering, University of Abomey-Calavi, Abomey-Calavi, Bénin

2 Experimental and Clinical Biology Laboratory, National School of Applied Biosciences and Biotechnologies, National University of Science, Technology, Engineering and Mathematics (UNSTIM), Dassa-Zoumé, Benin

*Address all correspondence to: gatien.lokossou@epac.uac.bj and gatien.lokossou@gmail.com

References

[1] Steegers EAP, von Dadelszen P, Duvekot JJ, Pijnenborg R. Pre-eclampsia. Lancet. 2010;**376**(9741):631-644

[2] Maher GM, O'Keeffe GW, Kenny LC, Kearney PM, Dinan TG, Khashan AS. Hypertensive disorders of pregnancy and risk of neurodevelopmental disorders in the offspring: A systematic review and meta-analysis protocol. BMJ Open. Oct 2017;7(10):e018313

[3] McLaren ZM, Sharp A, Hessburg JP, Sarvestani AS, Parker E, Akazili J, et al. Cost effectiveness of medical devices to diagnose pre-eclampsia in low-resource settings. Development Engineering. 2017;**2**:99-106

[4] von Dadelszen P, Magee LA. Pre-eclampsia: An Update. Current Hypertension Reports. 2014 [cité 28 déc 2017];**16**(8):454-468. DOI: 10.1007/s11906-014-0454-8

[5] Chaiworapongsa T, Chaemsaithong P, Yeo L, Romero R. Pre-eclampsia part 1: Current understanding of its pathophysiology. Nature Reviews Nephrology. 2014;**10**(8):466-480

[6] Hsu P, Nanan RKH. Innate and adaptive immune interactions at the fetal and maternal Interface in healthy human pregnancy and pre-eclampsia. Frontiers in Immunology. 2014 [cité 26 janv 2018];**5**:1-12. DOI: 10.3389/fimmu.2014.00125/abstract

[7] Luppi P, Tse H, Lain KY, Markovic N, Piganelli JD, DeLoia JA. Preeclampsia activates circulating immune cells with engagement of the NF-kappaB pathway. American Journal of Reproductive Immunology. 2006;**56**(2):135-144

[8] Luppi P, DeLoia JA. Monocytes of preeclamptic women spontaneously synthesize pro-inflammatory cytokines. Clinical Immunology. 2006;**118**(2-3):268-275

[9] Muzzio DO, Soldati R, Ehrhardt J, Utpatel K, Evert M, Zenclussen AC, et al. B cell development undergoes profound modifications and adaptations during pregnancy in Mice1. Biology of Reproduction. 2014 [cité 7 avr 2022];**91**(5):29-37. DOI: 10.1095/biolreprod.114.122366

[10] Guzman-Genuino RM, Diener KR. Regulatory B cells in pregnancy: Lessons from autoimmunity, graft tolerance, and Cancer. Frontiers in Immunology. 2017 [cité 7 avr 2022];**8**:1-19. DOI: 10.3389/fimmu.2017.00172/full

[11] Magatti M, Masserdotti A, Cargnoni A, Papait A, Stefani FR, Silini AR, et al. The role of B cells in PE pathophysiology: A potential target for perinatal cell-based therapy? IJMS. 2021;**22**(7):3405

[12] Dar HY, Rani L, Sapra L, Azam Z, Shokeen N, Bhardwaj A, et al. A chronological journey of Breg subsets: Implications in health and disease. In: Singh S, editor. Systems and Synthetic Immunology. Singapore: Springer Singapore; 2020 [cité 8 sept 2022]. pp. 125-152. DOI: 10.1007/978-981-15-3350-1_5

[13] Busse M, Campe KNJ, Redlich A, Oettel A, Hartig R, Costa SD, et al. Regulatory B cells are decreased and impaired in their function in peripheral maternal blood in pre-term birth. Frontiers in Immunology. 2020;**11**:386

[14] Boldison J, Da Rosa LC, Davies J, Wen L, Wong FS. Dendritic cells license regulatory B cells to produce IL-10 and

mediate suppression of antigen-specific CD8 T cells. Cellular & Molecular Immunology. 2020;**17**(8):843-855

[15] Busse M, Campe KNJ, Nowak D, Schumacher A, Plenagl S, Langwisch S, et al. IL-10 producing B cells rescue mouse fetuses from inflammation-driven fetal death and are able to modulate T cell immune responses. Scientific Reports. 2019;**9**(1):9335

[16] Rolle L, Memarzadeh Tehran M, Morell-García A, Raeva Y, Schumacher A, Hartig R, et al. Cutting edge: IL-10-producing regulatory B cells in early human pregnancy. American Journal of Reproductive Immunology. 2013;**70**(6):448-453

[17] Jensen F, Muzzio D, Soldati R, Fest S, Zenclussen AC. Regulatory B10 cells restore pregnancy tolerance in a mouse Model1. Biology of Reproduction. 2013 [cité 20 mai 2020];**89**(4):1-7. DOI: 10.1095/biolreprod.113.110791

[18] King A, Burrows TD, Hiby SE, Bowen JM, Joseph S, Verma S, et al. Surface expression of HLA-C antigen by human extravillous trophoblast. Placenta. 2000;**21**(4):376-387

[19] Ishitani A, Sageshima N, Lee N, Dorofeeva N, Hatake K, Marquardt H, et al. Protein expression and peptide binding suggest unique and interacting functional roles for HLA-E, F, and G in maternal-placental immune recognition. Journal of Immunology. 2003;**171**(3):1376-1384

[20] Kovats S, Main EK, Librach C, Stubblebine M, Fisher SJ, DeMars R. A class I antigen, HLA-G, expressed in human trophoblasts. Science. 1990;**248**(4952):220-223

[21] Segerer SE, Staib C, Kaemmerer U, Frambach T, Honig A, Dietl J, et al. Dendritic cells: Elegant arbiters in human reproduction. Current Pharmaceutical Biotechnology. 2012;**13**(8):1378-1384

[22] Ruocco MG, Chaouat G, Florez L, Bensussan A, Klatzmann D. Regulatory T-cells in pregnancy: Historical perspective, state of the art, and burning questions. Frontiers in Immunology. 2014 [cité 8 sept 2022];**5**:1-10. DOI: 10.3389/fimmu.2014.00389/abstract

[23] Fillatreau S, Sweenie CH, McGeachy MJ, Gray D, Anderton SM. B cells regulate autoimmunity by provision of IL-10. Nature Immunology. 2002;**3**(10):944-950

[24] Matsushita T, Yanaba K, Bouaziz JD, Fujimoto M, Tedder TF. Regulatory B cells inhibit EAE initiation in mice while other B cells promote disease progression. The Journal of Clinical Investigation. 2008;**118**(10):3420-3430

[25] Newell KA, Asare A, Kirk AD, Gisler TD, Bourcier K, Suthanthiran M, et al. Identification of a B cell signature associated with renal transplant tolerance in humans. Journal of Clinical Investigation. 2010;**120**(6):1836-1847

[26] Mauri C, Menon M. The expanding family of regulatory B cells. International Immunology. 2015;**27**(10):479-486

[27] Rosser EC, Mauri C. Regulatory B cells: Origin, phenotype, and function. Immunity. 2015;**42**(4):607-612

[28] Yanaba K, Bouaziz JD, Haas KM, Poe JC, Fujimoto M, Tedder TF. A regulatory B cell subset with a unique CD1dhiCD5+ phenotype controls T cell-dependent inflammatory responses. Immunity. 2008;**28**(5):639-650

[29] Mauri C, Gray D, Mushtaq N, Londei M. Prevention of arthritis by interleukin 10-producing B cells. The

Journal of Experimental Medicine. 2003;**197**(4):489-501

[30] Blair LYN, Flores-Borja F, Rawlings DJ, Isenberg DA, Ehrenstein MR, et al. CD19(+)CD24(hi) CD38(hi) B cells exhibit regulatory capacity in healthy individuals but are functionally impaired in systemic lupus erythematosus patients. 2010;**32**(1):129-140. DOI: 10.1016/j.immuni.2009.11.009

[31] Hamilton WJ, Boyd JD. Development of the human placenta in the first three months of gestation. Journal of Anatomy. 1960;**94**:297-328

[32] Castellucci M, Scheper M, Scheffen I, Celona A, Kaufmann P. The development of the human placental villous tree. Anatomy and Embryology (Berl). 1990;**181**(2):117-128

[33] Jones CJ, Fox H. Ultrastructure of the normal human placenta. Electron Microscopy Reviews. 1991;**4**(1):129-178

[34] Fisher SJ. Why is placentation abnormal in preeclampsia? American Journal of Obstetrics and Gynecology. 2015;**213**(4 Suppl):S115-S122

[35] Khong TY, Liddell HS, Robertson WB. Defective haemochorial placentation as a cause of miscarriage: A preliminary study. British Journal of Obstetrics and Gynaecology. 1987;**94**(7):649-655

[36] Khong TY, De Wolf F, Robertson WB, Brosens I. Inadequate maternal vascular response to placentation in pregnancies complicated by pre-eclampsia and by small-for-gestational age infants. British Journal of Obstetrics and Gynaecology. 1986;**93**(10):1049-1059

[37] Gude NM, Roberts CT, Kalionis B, King RG. Growth and function of the normal human placenta. Thrombosis Research. 2004;**114**(5-6):397-407

[38] Xiao Z, Yan L, Liang X, Wang H. Progress in deciphering trophoblast cell differentiation during human placentation. Current Opinion in Cell Biology. 2020;**67**:86-91

[39] Soilleux EJ, Coleman N. Transplacental transmission of HIV: A potential role for HIV binding lectins. The International Journal of Biochemistry & Cell Biology. 2003;**35**(3):283-287

[40] Arechavaleta-Velasco F, Koi H, Strauss JF, Parry S. Viral infection of the trophoblast: Time to take a serious look at its role in abnormal implantation and placentation? Journal of Reproductive Immunology. 2002;**55**(1-2):113-121

[41] Burton GJ, Jauniaux E. The cytotrophoblastic shell and complications of pregnancy. Placenta. 2017;**60**:134-139

[42] Velicky P, Meinhardt G, Plessl K, Vondra S, Weiss T, Haslinger P, et al. Genome amplification and cellular senescence are hallmarks of human placenta development. PLoS Genetics. 2018;**14**(10):e1007698

[43] Pijnenborg R, Dixon G, Robertson WB, Brosens I. Trophoblastic invasion of human decidua from 8 to 18 weeks of pregnancy. Placenta. 1980;**1**(1):3-19

[44] Baczyk D, Drewlo S, Proctor L, Dunk C, Lye S, Kingdom J. Glial cell missing-1 transcription factor is required for the differentiation of the human trophoblast. Cell Death and Differentiation. 2009;**16**(5):719-727

[45] Renaud SJ, Kubota K, Rumi MAK, Soares MJ. The FOS transcription factor family differentially controls

trophoblast migration and invasion. The Journal of Biological Chemistry. 2014;**289**(8):5025-5039

[46] Gatien L. Low serum-derived Syncytin-2 levels in exosome at early pregnancy is a predictor of preeclampsia: A prospective pilot study in Benin, West Africa. Journal of Obstetrics and Gynecological Surgery. 2020;**17**:1-6

[47] Nguyen TPH, Patrick CJ, Parry LJ, Familari M. Using proteomics to advance the search for potential biomarkers for preeclampsia: A systematic review and meta-analysis. PLoS One. 2019;**14**(4):e0214671

[48] Khaliq OP, Konoshita T, Moodley J, Naicker T. The role of uric acid in preeclampsia: Is uric acid a causative factor or a sign of preeclampsia? Current Hypertension Reports. 2018;**20**(9):80

[49] Lokossou AG, Toufaily C, Vargas A, Barbeau B. siRNA transfection and EMSA analyses on freshly isolated human villous Cytotrophoblasts. Journal of Visualized Experiments. 2016;**115**:1-9

[50] Vargas A, Moreau J, Landry S, LeBellego F, Toufaily C, Rassart É, et al. Syncytin-2 plays an important role in the fusion of human trophoblast cells. Journal of Molecular Biology. 2009;**392**(2):301-318

[51] Langbein M, Strick R, Strissel PL, Vogt N, Parsch H, Beckmann MW, et al. Impaired cytotrophoblast cell-cell fusion is associated with reduced Syncytin and increased apoptosis in patients with placental dysfunction. Molecular Reproduction and Development. 2008;**75**(1):175-183

[52] Huppertz B. Placental origins of preeclampsia: Challenging the current hypothesis. Hypertension. 2008;**51**(4):970-975

[53] Vargas A, Toufaily C, LeBellego F, Rassart É, Lafond J, Barbeau B. Reduced expression of both Syncytin 1 and Syncytin 2 correlates with severity of preeclampsia. Reproductive Sciences. 2011;**18**(11):1085-1091

[54] Lu X, Wang R, Zhu C, Wang H, Lin HY, Gu Y, et al. Fine-tuned and cell-cycle-restricted expression of Fusogenic protein Syncytin-2 maintains functional placental syncytia. Cell Reports. 2017;**21**(5):1150-1159

[55] Vargas A, Zhou S, Éthier-Chiasson M, Flipo D, Lafond J, Gilbert C, et al. Syncytin proteins incorporated in placenta exosomes are important for cell uptake and show variation in abundance in serum exosomes from patients with preeclampsia. The FASEB Journal. 2014;**28**(8):3703-3719

[56] Lokossou AG, Toudic C, Nguyen PT, Elisseeff X, Vargas A, Rassart É, et al. Endogenous retrovirus-encoded Syncytin-2 contributes to exosome-mediated immunosuppression of T cells. Biology of Reproduction. 2020;**102**(1):185-198

[57] Matsubara K, Matsubara Y, Uchikura Y, Sugiyama T. Pathophysiology of preeclampsia: The role of exosomes. International Journal of Molecular Sciences. 2021;**22**(5):2572

[58] Levine RJ, Maynard SE, Qian C, Lim KH, England LJ, Yu KF, et al. Circulating angiogenic factors and the risk of preeclampsia. The New England Journal of Medicine. 2004;**350**(7):672-683

[59] Bobrie A, Colombo M, Raposo G, Théry C. Exosome secretion: Molecular mechanisms and roles in immune responses. Traffic. 2011;**12**(12):1659-1668

[60] Lokossou AG, Toudic C, Barbeau B. Implication of human

endogenous retrovirus envelope proteins in placental functions. Viruses. 2014;**6**(11):4609-4627

[61] Kar M. Role of biomarkers in early detection of preeclampsia. Journal of Clinical and Diagnostic Research. 2014;**8**(4):BE01-BE04

[62] Valadi H, Ekström K, Bossios A, Sjöstrand M, Lee JJ, Lötvall JO. Exosome-mediated transfer of mRNAs and microRNAs is a novel mechanism of genetic exchange between cells. Nature Cell Biology. 2007;**9**(6):654-659

[63] Konečná B, Tóthová Ľ, Repiská G. Exosomes-associated DNA-new marker in pregnancy complications? International Journal of Molecular Sciences. 2019;**20**(12):E2890

[64] Bowers EC, Hassanin AAI, Ramos KS. In vitro models of exosome biology and toxicology: New frontiers in biomedical research. Toxicology in Vitro. 2020;**64**:104462

[65] Thomou T, Mori MA, Dreyfuss JM, Konishi M, Sakaguchi M, Wolfrum C, et al. Adipose-derived circulating miRNAs regulate gene expression in other tissues. Nature. 2017;**542**(7642):450-455

[66] Czernek L, Düchler M. Exosomes as messengers between mother and fetus in pregnancy. International Journal of Molecular Sciences. 2020;**21**(12):E4264

[67] Pillay P, Moodley K, Moodley J, Mackraj I. Placenta-derived exosomes: Potential biomarkers of preeclampsia. International Journal of Nanomedicine. 2017;**12**:8009-8023

[68] Devor E, Santillan D, Scroggins S, Warrier A, Santillan M. Trimester-specific plasma exosome microRNA expression profiles in preeclampsia. The Journal of Maternal-Fetal & Neonatal Medicine. 2020;**33**(18):3116-3124

[69] Gill M, Motta-Mejia C, Kandzija N, Cooke W, Zhang W, Cerdeira AS, et al. Placental Syncytiotrophoblast-derived extracellular vesicles carry active NEP (Neprilysin) and are increased in preeclampsia. Hypertension. 2019;**73**(5):1112-1119

[70] Kshirsagar SK, Alam SM, Jasti S, Hodes H, Nauser T, Gilliam M, et al. Immunomodulatory molecules are released from the first trimester and term placenta via exosomes. Placenta. 2012;**33**(12):982-990

[71] Mitchell MD, Peiris HN, Kobayashi M, Koh YQ, Duncombe G, Illanes SE, et al. Placental exosomes in normal and complicated pregnancy. American Journal of Obstetrics and Gynecology. 2015;**213**(4):S173-S181

[72] Pillay P, Maharaj N, Moodley J, Mackraj I. Placental exosomes and pre-eclampsia: Maternal circulating levels in normal pregnancies and, early and late onset pre-eclamptic pregnancies. Placenta. 2016;**46**:18-25

[73] Saito S, Sakai M. Th1/Th2 balance in preeclampsia. Journal of Reproductive Immunology. 2003;**59**(2):161-173

[74] Toufaily C, Lokossou AG, Vargas A, Rassart É, Barbeau B. A CRE/AP-1-like motif is essential for induced syncytin-2 expression and fusion in human trophoblast-like model. PLoS One. 2015;**10**(3):e0121468

[75] Ghaebi M, Nouri M, Ghasemzadeh A, Farzadi L, Jadidi-Niaragh F, Ahmadi M, et al. Immune regulatory network in successful pregnancy and reproductive failures. Biomedicine & Pharmacotherapy. 2017;**88**:61-73

[76] Miller D, Motomura K, Galaz J, Gershater M, Lee ED, Romero R, et al. Cellular immune responses in the pathophysiology of preeclampsia. Journal of Leukocyte Biology. 2022;**111**(1):237-260

[77] Chappell LC, Enye S, Seed P, Briley AL, Poston L, Shennan AH. Adverse perinatal outcomes and risk factors for preeclampsia in women with chronic hypertension: A prospective study. Hypertension. 2008;**51**(4):1002-1009

[78] Roberts JM, Taylor RN, Musci TJ, Rodgers GM, Hubel CA, McLaughlin MK. Preeclampsia: An endothelial cell disorder. American Journal of Obstetrics and Gynecology. 1989;**161**(5):1200-1204

[79] Roberts JM, Hubel CA. The two stage model of preeclampsia: Variations on the theme. Placenta. 2009;**30**(Suppl. A): S32-S37

[80] Soleymanlou N, Jurisica I, Nevo O, Ietta F, Zhang X, Zamudio S, et al. Molecular evidence of placental hypoxia in preeclampsia. The Journal of Clinical Endocrinology and Metabolism. 2005;**90**(7):4299-4308

[81] Abe E, Matsubara K, Ochi H, Ito M, Oka K, Kameda K. Elevated levels of adhesion molecules derived from leukocytes and endothelial cells in patients with pregnancy-induced hypertension. Hypertension in Pregnancy. 2003;**22**(1):31-43

[82] Abe E, Matsubara K, Oka K, Kusanagi Y, Ito M. Cytokine regulation of intercellular adhesion molecule-1 expression on trophoblasts in preeclampsia. Gynecologic and Obstetric Investigation. 2008;**66**(1):27-33

[83] Matsubara K, Abe E, Ochi H, Kusanagi Y, Ito M. Changes in serum concentrations of tumor necrosis factor alpha and adhesion molecules in normal pregnant women and those with pregnancy-induced hypertension. The Journal of Obstetrics and Gynaecology Research. 2003;**29**(6):422-426

[84] Ballegeer VC, Spitz B, De Baene LA, Van Assche AF, Hidajat M, Criel AM. Platelet activation and vascular damage in gestational hypertension. American Journal of Obstetrics and Gynecology. 1992;**166**(2):629-633

[85] Roberts JM, Taylor RN, Goldfien A. Clinical and biochemical evidence of endothelial cell dysfunction in the pregnancy syndrome preeclampsia. American Journal of Hypertension. 1991;**4**(8):700-708

[86] Tjoa ML, Jani J, Lewi L, Peter I, Wataganara T, Johnson KL, et al. Circulating cell-free fetal messenger RNA levels after fetoscopic interventions of complicated pregnancies. American Journal of Obstetrics and Gynecology. 2006;**195**(1):230-235

[87] Taglauer ES, Wilkins-Haug L, Bianchi DW. Review: Cell-free fetal DNA in the maternal circulation as an indication of placental health and disease. Placenta. 2014;**35**(Suppl):S64-S68

[88] Yanagisawa M, Kurihara H, Kimura S, Tomobe Y, Kobayashi M, Mitsui Y, et al. A novel potent vasoconstrictor peptide produced by vascular endothelial cells. Nature. 1988;**332**(6163):411-415

[89] Palmer RM, Ashton DS, Moncada S. Vascular endothelial cells synthesize nitric oxide from L-arginine. Nature. 1988;**333**(6174):664-666

[90] Roberts JM, Hubel CA. Is oxidative stress the link in the two-stage

model of pre-eclampsia? Lancet. 1999;**354**(9181):788-789

[91] Matsubara K, Matsubara Y, Hyodo S, Katayama T, Ito M. Role of nitric oxide and reactive oxygen species in the pathogenesis of preeclampsia. The Journal of Obstetrics and Gynaecology Research. 2010;**36**(2):239-247

[92] Kifle MM, Dahal P, Vatish M, Cerdeira AS, Ohuma EO. The prognostic utility of soluble fms-like tyrosine kinase-1 (sFlt-1) and placental growth factor (PIGF) biomarkers for predicting preeclampsia: A secondary analysis of data from the INSPIRE trial. BMC Pregnancy and Childbirth. 2022;**22**(1):520

[93] Burton GJ, Jauniaux E. Placental oxidative stress: From miscarriage to preeclampsia. Journal of the Society for Gynecologic Investigation. 2004;**11**(6):342-352

[94] Schlembach D, Hund M, Schroer A, Wolf C. Economic assessment of the use of the sFlt-1/PlGF ratio test to predict preeclampsia in Germany. BMC Health Services Research. 6 août 2018;**18**(1):603

[95] Hodel M, Blank PR, Marty P, Lapaire O. sFlt-1/PlGF ratio as a predictive marker in women with suspected preeclampsia: An economic evaluation from a Swiss perspective. Disease Markers. 2019;**2019**:4096847

[96] Brown MA, Lindheimer MD, de Swiet M, Van Assche A, Moutquin JM. The classification and diagnosis of the hypertensive disorders of pregnancy: Statement from the International Society for the Study of hypertension in pregnancy (ISSHP). Hypertension in Pregnancy. 2001;**20**(1):IX-XIV

[97] Phipps E, Prasanna D, Brima W, Jim B. Preeclampsia: Updates in pathogenesis, definitions, and guidelines. Clinical Journal of the American Society of Nephrology. 2016;**11**(6):1102-1113

[98] Robillard PY, Hulsey TC, Alexander GR, Keenan A, de Caunes F, Papiernik E. Paternity patterns and risk of preeclampsia in the last pregnancy in multiparae. Journal of Reproductive Immunology. 1993;**24**(1):1-12

[99] Skjaerven R, Wilcox AJ, Lie RT. The interval between pregnancies and the risk of preeclampsia. The New England Journal of Medicine. 2002;**346**(1):33-38

[100] Salha O, Sharma V, Dada T, Nugent D, Rutherford AJ, Tomlinson AJ, et al. The influence of donated gametes on the incidence of hypertensive disorders of pregnancy. Human Reproduction. 1999;**14**(9):2268-2273

[101] Melchiorre K, Giorgione V, Thilaganathan B. The placenta and preeclampsia: Villain or victim? American Journal of Obstetrics and Gynecology. 2022;**226**(2S):S954-S962

[102] Klonoff-Cohen HS, Savitz DA, Cefalo RC, McCann MF. An epidemiologic study of contraception and preeclampsia. Journal of the American Medical Association. 1989;**262**(22):3143-3147

[103] Koelman CA, Coumans AB, Nijman HW, Doxiadis II, Dekker GA, Claas FH. Correlation between oral sex and a low incidence of preeclampsia: A role for soluble HLA in seminal fluid? Journal of Reproductive Immunology. 2000;**46**(2):155-166

[104] Billingham RE, Brent L, Medawar PB. Actively acquired tolerance of foreign cells. Nature. 1953;**172**(4379): 603-606

[105] Saito S, Nishikawa K, Morii T, Narita N, Enomoto M, Ichijo M.

Expression of activation antigens CD69, HLA-DR, interleukin-2 receptor-alpha (IL-2R alpha) and IL-2R beta on T cells of human decidua at an early stage of pregnancy. Immunology. 1992;**75**(4):710-712

[106] Aluvihare VR, Kallikourdis M, Betz AG. Regulatory T cells mediate maternal tolerance to the fetus. Nature Immunology. 2004;**5**(3):266-271

[107] Saito S, Shiozaki A, Nakashima A, Sakai M, Sasaki Y. The role of the immune system in preeclampsia. Molecular Aspects of Medicine. 2007;**28**(2):192-209

[108] Nagamatsu T, Schust DJ, Sugimoto J, Barrier BF. Human decidual stromal cells suppress cytokine secretion by allogenic CD4+ T cells via PD-1 ligand interactions. Human Reproduction. 2009;**24**(12):3160-3171

[109] Freeman GJ, Long AJ, Iwai Y, Bourque K, Chernova T, Nishimura H, et al. Engagement of the PD-1 immunoinhibitory receptor by a novel B7 family member leads to negative regulation of lymphocyte activation. The Journal of Experimental Medicine. 2000;**192**(7):1027-1034

[110] Munn DH, Zhou M, Attwood JT, Bondarev I, Conway SJ, Marshall B, et al. Prevention of allogeneic fetal rejection by tryptophan catabolism. Science. 1998;**281**(5380):1191-1193

[111] Koopman LA, Kopcow HD, Rybalov B, Boyson JE, Orange JS, Schatz F, et al. Human Decidual natural killer cells are a unique NK cell subset with immunomodulatory potential. Journal of Experimental Medicine. 2003;**198**(8):1201-1212

[112] Higuma-Myojo S, Sasaki Y, Miyazaki S, Sakai M, Siozaki A, Miwa N, et al. Cytokine profile of natural killer cells in early human pregnancy. American Journal of Reproductive Immunology. 2005;**54**(1):21-29

[113] Darmochwal-Kolarz D, Kolarz B, Rolinski J, Leszczynska-Gorzelak B, Oleszczuk J. The concentrations of soluble HLA-G protein are elevated during mid-gestation and decreased in pre-eclampsia. Folia Histochemica et Cytobiologica. 2012;**50**(2):286-291

[114] Xu X, Zhou Y, Wei H. Roles of HLA-G in the maternal-fetal immune microenvironment. Frontiers in Immunology. 2020;**11**:592010

[115] Saito S, Shima T, Nakashima A, Inada K, Yoshino O. Role of paternal antigen-specific Treg cells in successful implantation. American Journal of Reproductive Immunology. 2016;**75**(3):310-316

[116] Murata H, Tanaka S, Tsuzuki-Nakao T, Kido T, Kakita-Kobayashi M, Kida N, et al. The transcription factor HAND2 up-regulates transcription of the IL15 gene in human endometrial stromal cells. The Journal of Biological Chemistry. 2020;**295**(28):9596-9605

[117] van der Zwan A, Bi K, Norwitz ER, Crespo ÂC, Claas FHJ, Strominger JL, et al. Mixed signature of activation and dysfunction allows human decidual CD8+ T cells to provide both tolerance and immunity. Proceedings of the National Academy of Sciences of the United States of America. 2018;**115**(2):385-390

[118] Tilburgs T, Schonkeren D, Eikmans M, Nagtzaam NM, Datema G, Swings GM, et al. Human decidual tissue contains differentiated CD8+ effector-memory T cells with unique properties. Journal of Immunology. 2010;**185**(7):4470-4477

[119] Morita K, Tsuda S, Kobayashi E, Hamana H, Tsuda K, Shima T, et al. Analysis of TCR repertoire and PD-1 expression in Decidual and peripheral CD8+ T cells reveals distinct immune mechanisms in miscarriage and preeclampsia. Frontiers in Immunology. 2020;**11**:1082

[120] Kinder JM, Turner LH, Stelzer IA, Miller-Handley H, Burg A, Shao TY, et al. CD8+ T cell functional exhaustion overrides pregnancy-induced fetal antigen Alloimmunization. Cell Reports. 2020;**31**(12):107784

[121] Tsuda S, Nakashima A, Morita K, Shima T, Yoneda S, Kishi H, et al. The role of decidual regulatory T cells in the induction and maintenance of fetal antigen-specific tolerance: Imbalance between regulatory and cytotoxic T cells in pregnancy complications. Human Immunology. 2021;**82**(5):346-352

[122] Robertson SA, Prins JR, Sharkey DJ, Moldenhauer LM. Seminal fluid and the generation of regulatory T cells for embryo implantation. American Journal of Reproductive Immunology. 2013;**69**(4):315-330

[123] Robertson SA, Guerin LR, Bromfield JJ, Branson KM, Ahlström AC, Care AS. Seminal fluid drives expansion of the CD4+CD25+ T regulatory cell pool and induces tolerance to paternal alloantigens in mice. Biology of Reproduction. 2009;**80**(5):1036-1045

[124] Shima T, Inada K, Nakashima A, Ushijima A, Ito M, Yoshino O, et al. Paternal antigen-specific proliferating regulatory T cells are increased in uterine-draining lymph nodes just before implantation and in pregnant uterus just after implantation by seminal plasma-priming in allogeneic mouse pregnancy. Journal of Reproductive Immunology. 2015;**108**:72-82

[125] Shima T, Nakashima A, Yasuda I, Ushijima A, Inada K, Tsuda S, et al. Uterine CD11c+ cells induce the development of paternal antigen-specific Tregs via seminal plasma priming. Journal of Reproductive Immunology. 2020;**141**:103165

[126] Yasuda I, Shima T, Moriya T, Ikebuchi R, Kusumoto Y, Ushijima A, et al. Dynamic changes in the phenotype of dendritic cells in the uterus and uterine draining lymph nodes after coitus. Frontiers in Immunology. 2020;**11**:557720

[127] Ono Y, Yoshino O, Hiraoka T, Sato E, Fukui Y, Ushijima A, et al. CD206+ M2-like macrophages are essential for successful implantation. Frontiers in Immunology. 2020;**11**:557184

[128] Salvany-Celades M, van der Zwan A, Benner M, Setrajcic-Dragos V, Bougleux Gomes HA, Iyer V, et al. Three types of functional regulatory T cells control T cell responses at the human maternal-fetal Interface. Cell Reports. 2019;**27**(9):2537-2547.e5

[129] Papuchova H, Kshirsagar S, Xu L, Bougleux Gomes HA, Li Q, Iyer V, et al. Three types of HLA-G+ extravillous trophoblasts that have distinct immune regulatory properties. Proceedings of the National Academy of Sciences of the United States of America. 2020;**117**(27):15772-15777

[130] Hosseini Teshnizi S, Ali-Hassanzadeh M, Gharesi-Fard B, Kabelitz D, Kalantar K. Influence of forkhead box protein 3 polymorphisms (rs2232365, rs3761548) with the outcome of pregnancy: A meta-analysis. Journal Cellular Physiology. 2019;**234**(9):16573-16581

[131] Tsuda S, Zhang X, Hamana H, Shima T, Ushijima A, Tsuda K, et al. Clonally expanded Decidual effector regulatory T cells increase in late

gestation of Normal pregnancy, but not in preeclampsia, in humans. Frontiers in Immunology. 2018;**9**:1934

[132] Vince GS, Starkey PM, Austgulen R, Kwiatkowski D, Redman CW. Interleukin-6, tumour necrosis factor and soluble tumour necrosis factor receptors in women with pre-eclampsia. British Journal of Obstetrics and Gynaecology. 1995;**102**(1):20-25

[133] Schumacher A, Costa SD, Zenclussen AC. Endocrine factors modulating immune responses in pregnancy. Frontiers in Immunology. 2014 [cité 9 sept 2022];**5**:1-12. DOI: 10.3389/fimmu.2014.00196/abstract

[134] Mizoguchi A, Mizoguchi E, Takedatsu H, Blumberg RS, Bhan AK. Chronic intestinal inflammatory condition generates IL-10-producing regulatory B cell subset characterized by CD1d upregulation. Immunity. 2002;**16**(2):219-230

[135] Robertson SA, Care AS, Skinner RJ. Interleukin 10 regulates inflammatory cytokine synthesis to protect against lipopolysaccharide-induced abortion and fetal growth restriction in Mice1. Biology of Reproduction. 2007;**76**(5):738-748

[136] Murphy SP, Fast LD, Hanna NN, Sharma S. Uterine NK cells mediate inflammation-induced fetal demise in IL-10-null mice. Journal of Immunology. 2005;**175**(6):4084-4090

[137] White CA, Johansson M, Roberts CT, Ramsay AJ, Robertson SA. Effect of Interleukin-10 null mutation on maternal immune response and reproductive outcome in Mice1. Biology of Reproduction. 2004;**70**(1):123-131

[138] Busse M, Redlich A, Hartig R, Costa SD, Rathert H, Fest S, et al. Imbalance between inflammatory and regulatory cord blood B cells following pre-term birth. Journal of Reproductive Immunology. 2021;**145**:103319

[139] Guzman-Genuino RM, Hayball JD, Diener KR. Regulatory B cells: Dark horse in pregnancy immunotherapy? Journal of Molecular Biology. 2021;**433**(1):166596

[140] Zenclussen AC. A novel mouse model for preeclampsia by transferring activated Th1 cells into Normal pregnant mice. In: Placenta and Trophoblast. New Jersey: Humana Press; 2005 [cité 9 sept 2022]. pp. 401-412. DOI: 10.1385/1-59259-989-3:401

[141] Zenclussen AC, Fest S, Joachim R, Klapp BF, Arck PC. Introducing a mouse model for pre-eclampsia: Adoptive transfer of activated Th1 cells leads to pre-eclampsia-like symptoms exclusively in pregnant mice. European Journal of Immunology. 2004;**34**(2):377-387

[142] Muzzio DO, Ziegler KB, Ehrhardt J, Zygmunt M, Jensen F. Marginal zone B cells emerge as a critical component of pregnancy well-being. Reproduction. 2016;**151**(1):29-37

[143] Morandi F, Horenstein AL, Quarona V, Faini AC, Castella B, Srinivasan RC, et al. Ectonucleotidase expression on human amnion epithelial cells: Adenosinergic pathways and Dichotomic effects on immune effector cell populations. Journal of Immunology. 2019;**202**(3):724-735

[144] Ribeiro A, Laranjeira P, Mendes S, Velada I, Leite C, Andrade P, et al. Mesenchymal stem cells from umbilical cord matrix, adipose tissue and bone marrow exhibit different capability to suppress peripheral blood B, natural killer and T cells. Stem Cell Research & Therapy. 2013;**4**(5):125

[145] Ji YR, Yang ZX, Han ZB, Meng L, Liang L, Feng XM, et al. Mesenchymal

stem cells support proliferation and terminal differentiation of B cells. Cellular Physiology and Biochemistry. 2012;**30**(6):1526-1537

[146] Silini AR, Di Pietro R, Lang-Olip I, Alviano F, Banerjee A, Basile M, et al. Perinatal derivatives: Where do we stand? A roadmap of the human placenta and consensus for tissue and cell nomenclature. Frontiers in Bioengineering and Biotechnology. 2020;**8**:610544

[147] Che N, Li X, Zhou S, Liu R, Shi D, Lu L, et al. Umbilical cord mesenchymal stem cells suppress B-cell proliferation and differentiation. Cellular Immunology. 2012;**274**(1-2):46-53

[148] Lin HD, Fong CY, Biswas A, Choolani M, Bongso A. Human Wharton's jelly stem cells, its conditioned medium and cell-free lysate inhibit the growth of human lymphoma cells. Stem Cell Reviews and Reports. 2014;**10**(4):573-586

[149] Ma L, Zhou Z, Zhang D, Yang S, Wang J, Xue F, et al. Immunosuppressive function of mesenchymal stem cells from human umbilical cord matrix in immune thrombocytopenia patients. Thrombosis and Haemostasis. 2012;**107**(5):937-950

[150] Xue Q, Yin Z, Varshithreddy N, Liang H s, Wang M y, Dong W l, et al. The immunomodulatory function of human amniotic fluid stromal cells on B lymphocytes. Journal of Neurorestoratology. 2018;**6**(1):122-133

[151] Jensen F, Wallukat G, Herse F, Budner O, El-Mousleh T, Costa SD, et al. CD19+CD5+ cells as indicators of preeclampsia. Hypertension. 2012;**59**(4):861-868

[152] Eledel R, Bassuoni M, Radwan W, Masoud A, Eldeeb S. CD19 $^+$ CD5 $^+$ B-cell expansion and risk of pre-eclampsia. Menoufia Medical Journal. 2016;**29**(2):319

[153] Wallukat G, Homuth V, Fischer T, Lindschau C, Horstkamp B, Jüpner A, et al. Patients with preeclampsia develop agonistic autoantibodies against the angiotensin AT1 receptor. The Journal of Clinical Investigation. 1999;**103**(7):945-952

[154] Silini AR, Masserdotti A, Papait A, Parolini O. Shaping the future of perinatal cells: Lessons from the past and interpretations of the present. Frontiers in Bioengineering and Biotechnology. 2019;**7**:75

[155] Chen K, Wang D, Du WT, Han ZB, Ren H, Chi Y, et al. Human umbilical cord mesenchymal stem cells hUC-MSCs exert immunosuppressive activities through a PGE2-dependent mechanism. Clinical Immunology. 2010;**135**(3):448-458

[156] Rossi D, Pianta S, Magatti M, Sedlmayr P, Parolini O. Characterization of the conditioned medium from amniotic membrane cells: Prostaglandins as key effectors of its immunomodulatory activity. PLoS One. 2012;**7**(10):e46956

[157] Magatti M, Vertua E, De Munari S, Caro M, Caruso M, Silini A, et al. Human amnion favours tissue repair by inducing the M1-to-M2 switch and enhancing M2 macrophage features. Journal of Tissue Engineering and Regenerative Medicine. 2017;**11**(10):2895-2911

[158] Cargnoni A, Ressel L, Rossi D, Poli A, Arienti D, Lombardi G, et al. Conditioned medium from amniotic mesenchymal tissue cells reduces progression of bleomycin-induced lung fibrosis. Cytotherapy. 2012;**14**(2):153-161

[159] Carbone A, Castellani S, Favia M, Diana A, Paracchini V, Di Gioia S, et al.

Correction of defective CFTR/ENaC function and tightness of cystic fibrosis airway epithelium by amniotic mesenchymal stromal (stem) cells. Journal of Cellular and Molecular Medicine. 2014;**18**(8):1631-1643

[160] Lee PH, Tu CT, Hsiao CC, Tsai MS, Ho CM, Cheng NC, et al. Antifibrotic activity of human placental amnion membrane-derived CD34+ mesenchymal stem/progenitor cell transplantation in mice with Thioacetamide-induced liver injury. Stem Cells Translational Medicine. 2016;**5**(11):1473-1484

[161] Parolini O, Souza-Moreira L, O'Valle F, Magatti M, Hernandez-Cortes P, Gonzalez-Rey E, et al. Therapeutic effect of human amniotic membrane-derived cells on experimental arthritis and other inflammatory disorders: Effects of HAMCs on experimental arthritis. Arthritis & Rheumatology. 2014;**66**(2):327-339

[162] Tuca AC, Ertl J, Hingerl K, Pichlsberger M, Fuchs J, Wurzer P, et al. Comparison of Matrigel and Matriderm as a carrier for human amnion-derived mesenchymal stem cells in wound healing. Placenta. 2016;**48**:99-103

[163] Pischiutta F, Brunelli L, Romele P, Silini A, Sammali E, Paracchini L, et al. Protection of brain injury by amniotic mesenchymal stromal cell-secreted metabolites. Critical Care Medicine. 2016;**44**(11):e1118-e1131

[164] Lin YC, Ko TL, Shih YH, Lin MYA, Fu TW, Hsiao HS, et al. Human umbilical mesenchymal stem cells promote recovery after ischemic stroke. Stroke. 2011;**42**(7):2045-2053

[165] Giampà C, Alvino A, Magatti M, Silini AR, Cardinale A, Paldino E, et al. Conditioned medium from amniotic cells protects striatal degeneration and ameliorates motor deficits in the R6/2 mouse model of Huntington's disease. Journal of Cellular and Molecular Medicine. 2019;**23**(2):1581-1592

[166] Wang H, Qiu X, Ni P, Qiu X, Lin X, Wu W, et al. Immunological characteristics of human umbilical cord mesenchymal stem cells and the therapeutic effects of their transplantion on hyperglycemia in diabetic rats. International Journal of Molecular Medicine. 2014;**33**(2):263-270

[167] Tan B, Yuan W, Li J, Yang P, Ge Z, Liu J, et al. Therapeutic effect of human amniotic epithelial cells in murine models of Hashimoto's thyroiditis and systemic lupus erythematosus. Cytotherapy. 2018;**20**(10):1247-1258

[168] Gao L, Zhang Y, Hu B, Liu J, Kong P, Lou S, et al. Phase II multicenter, randomized, double-blind controlled study of efficacy and safety of umbilical cord-derived mesenchymal stromal cells in the prophylaxis of chronic graft-versus-host disease after HLA-Haploidentical stem-cell transplantation. Journal of Clinical Oncology. 2016;**34**(24):2843-2850

[169] Papait A, Vertua E, Magatti M, Ceccariglia S, De Munari S, Silini AR, et al. Mesenchymal stromal cells from fetal and maternal placenta possess key similarities and differences: Potential implications for their applications in regenerative medicine. Cell. 2020;**9**(1):E127

Chapter 8

Perspective Chapter: Engineering Secretory IgA against Infectious Diseases

Nicole Falci Finardi, Audrey Y-H. Teh and Julian K-C. Ma

Abstract

The dawn of antibody therapy was heralded by the rise of IgG therapeutics. However, other antibody classes are at our disposal—one of the most exciting is IgA and is the most abundant antibody class within humans. Unlike IgG, it is uniquely specialized for mucosal applications due to its ability to form complex Secretory IgA (SIgA) molecules. Since the mucosa is constantly exposed to potential infectious agents, SIgA is pivotal to disease prevention as an important component of the mucosal barrier. Compared to IgG, SIgA has proven superior effectiveness in mucosal surfaces, such as the airway epithelium or the harsh gut environment. Despite this, hurdles associated with low yield and challenging purification have blocked SIgA therapeutic advancement. However, as a result of new antibody engineering strategies, we are approaching the next generation of (IgA-based) antibody therapies. Strategies include fine-tuning SIgA assembly, exploring different production platforms, genetic engineering to improve purification, and glycoengineering of different components. Due to its stability in mucosal environments, SIgA therapeutics would revolutionize passive mucosal immunotherapy—an avenue still underexploited by current therapeutics. This chapter will focus on the current perspectives of SIgA engineering and explore different approaches to unlocking the full therapeutic potential of SIgAs.

Keywords: antibody engineering, infectious diseases, mucosal immunity, glycoengineering, mucosal antibodies, secretory IgA, IgA

1. Introduction

Out of all the antibody classes, the most abundant isoform in the human body is IgA [1]. It is the predominant antibody found in the mucosa—a vast extracellular environment constantly exposed to antigens from both pathogens and commensal bacteria—where it is secreted in the form of secretory IgA (SIgA). This is a multi-component molecule comprised of two IgA monomers linked together by a joining chain (J chain) and covalently attached to the secretory component (SC) [2]. The secretory component wraps around the antibody complex and confers resistance to proteolytic degradation, along with protection in low pH environments. The ability to form this stable structure is exclusive to mucosal antibodies IgA and IgM and presents a unique advantage to traditional IgG therapy which currently dominates the immunotherapy market [3]. Despite this, there are currently no licenced SIgA products available for use.

Being adapted to mucosal secretions means SIgA has an intrinsic advantage as an oral therapeutic targeting mucosal pathogens compared to other antibody classes. In airway infections, SIgA has been shown to neutralise influenza virus and prevent virus-induced pathology in the upper respiratory tract, at times better than IgG [4, 5]. This is likely to be due to its ability to bind antigens with high avidity on the mucosal surface and prevent adherence to the epithelium, a method called immune exclusion [3]. SIgA is anchored to the mucosal surface by interacting with mucins (glycoproteins which make up the mucus layer) via the antibody's secretory component, ensuring a layer of protection against potential pathogens [6]. More recently, the respiratory application of SIgA therapy was highlighted against COVID-19, with a monoclonal SIgA, but not IgG, potently neutralising SARS-CoV-2 virus [7]. This was attributed to the increased avidity of SIgA arising from its polymeric and flexible Fab regions.

SIgA is also a key immune component of gut mucosa, where it is secreted as the first line of defence [8]. This is an exceedingly complex environment where part of the interplay between host and gut microbiota can be disrupted by intestinal pathogens to establish infection (such as *Escherichia coli*, *Campylobacter jejuni* and *Clostridium difficile*) [9]. There is a global need to engineer therapeutics for diarrhoeal diseases which are a leading cause of death in developing countries, the majority being of young children [10]. In the search for therapeutics against such infections, SIgA has emerged as an attractive candidate. Administered via the oral route, SIgA against enterotoxigenic *E. coli* (ETEC) reduced instances of diarrheal disease by hindering bacterial adhesion to the gut lumen in a nonhuman primate model [11]. When compared to an IgG counterpart, SIgA neutralised a *C. difficile* toxin up to 100 times more effectively, although this was not replicated by all SIgA subtypes [12]. SIgA has also been demonstrated to be more stable in a low pH simulated intestinal fluid environment than the IgG1 variant [13]. The protective barrier arising from the interaction between SIgA and the mucus is effective at preventing bacteria from anchoring and colonising the gut lumen—the neutralised bacteria are then unable to cause disease.

With functional efficacy and applications against a wide range of mucosal pathogens, it is understandable why SIgA emerges as an attractive candidate in the growing field of antibody therapy. However, there are a host of factors which must be addressed before this antibody type can fulfil its therapeutic potential. Issues relating to SIgA production, purification and glycosylation must be addressed—the good news is there are already extensive efforts to do so.

2. Structure and assembly

SIgA is a complex molecule which relies on all 4 components assembling to form a functional antibody. This can be an issue when producing SIgA *in vitro* as partially assembled forms will also be produced (such as monomers and dimers). Furthermore, there are 2 subtypes of IgA which mainly differ in their hinge region: IgA1 with a long O-glycosylated hinge and IgA2 with a short non-glycosylated hinge [14]. Hinge length is associated with differences in flexibility and the ability to reach more distant epitopes at the expense of increased protease-mediated breakdown susceptibility [15]. As a result, the IgA subtype used for a specific therapeutic may be dependent on its application—for example, a potential SIgA therapeutic administered to the bacteria-rich gut mucosa might benefit from being IgA2 in order to evade hinge degradation by bacterial proteases. On the other hand, IgA1 may offer high stability due to the presence of strong covalent bonds between its heavy and light chains, otherwise absent in IgA2.

Engineering the IgA heavy chain in order to reduce sensitivity to bacterial proteases has highlighted which antibody domains are necessary for protease activity [13]. For example, three amino acids in the CH3 domain were found to confer susceptibility to breakdown by a *Neiseria meningitidis* protease. Efforts to engineer a IgA1/IgA2 hybrid with half of IgA1's long hinge has elucidated the specific proteases which bind to each half of the long hinge [16]. For example, proteases produced by Neisseria *gonorrhoeae* and *Neisseria meningitidis* are able to cleave the hybrid hinge, whereas the *Haemophilus influenzae Type 1* protease did not.

Protein structure can be stabilised by covalent bonds or non-covalent forces. A strategy in antibody engineering is introducing covalent bonds to antibody domains in the form of disulphide bridges. These make the antibody complex more stable and less prone to breakdown. For example, a single amino acid mutation (P221R) will sterically allow new covalent bonding in IgA2 between the heavy and light chain interaction—this leads to a more robust antibody and less free light chain [17].

The incorporation of J chain has been identified as a bottleneck in SIgA assembly [18]. This small 15 kDa polypeptide covalently attaches to the Fc region of opposite IgA monomers via two key cysteines to make dimeric IgA. However, excess J chain can lead to high molecular weight aggregation due to the protein's two free thiol groups [19]. Fine-tuning the expression of J chain relative to the other SIgA components, for example by putting it under the control of a stronger promoter, can help optimise dimeric and secretory forms of the antibody [20].

The final step to make a SIgA molecule is attachment of the secretory component. *In vivo* this happens as dimeric IgA is trancytosed from the lamina propria into mucosal secretions after binding to the polymeric immunoglobulin receptor [21]; *in cellulo* strategies rely on the SIgA complex assembling through co-expression of multiple genes or *in vitro* by incubating free secretory component with purified dimeric IgA [22].

There has been interest in engineering IgA fused with other immunoglobulin forms, particularly with camelid VHH nanobodies, replacing the variable light and heavy chains. These nanobodies are small (~15 kDa compared to the ~55 kDa Fab domain) and eliminate the need for a light chain. This enables the targeting of epitopes in deeper antigenic clefts and also simplifies production [23]. VHH-IgA fusions demonstrated increased functionality whilst retaining the ability to dimerise via the J chain and bind to secretory component. For example, secretory VHH-IgA fusions were protective against ETEC infection unlike related VHH-IgG fusions [24]. Another engineered fusion antibody which has been explored is an IgG backbone with IgA Fc sequences inserted—this demonstrated binding to both the IgA receptor (FcαRI) and IgG receptors (FcγRI/FcγRIIa/ FcγRIIb), possessing both classes of effector function [25].

With the recent increased application of advanced data science and machine learning in biotechnology, the coming years will potentially be very exciting for antibody engineering. Increased access to technology such as AlphaFold by Google, an open-source software to predict the 3D structure of protein sequences, will facilitate the bridging between *in vitro* and digital modelling [26]. Future approaches to improve antibody stability may employ computational models to identify antibody conformations with increased stability [27]. Indeed, proof-of-concept studies that apply different types of modelling software to antibody design are already underway, either by designing new proteins using an existing sequence dataset, or improving a 3D model and predicting the sequence which would give rise to it [28]. Such strategies have the potential to take native antibody sequences and generate novel sequences with better antigen binding affinity, for example [29]. It is important to note that this strategy requires a robust data set to base new designs on, and that improved antibody characteristics are based on simulation data which may not completely translate *in vitro/in vivo.* However, this is a relatively new technology. More experimental real-world data will both help guide and validate algorithms—unfortunately this information is sparse for IgA compared to IgG. Nevertheless, *in silico* antibody engineering has the power to drastically reduce the hours spent in laborious real-world antibody screening [30]. It is only a matter of when, not if, it will be applied to mainstream SIgA production.

3. Production: mammalian cells, plants and beyond

Production platforms for monoclonal antibodies determine the cost effectiveness and hence their viability as a therapeutic product. Due to the need to transcribe and assemble 4 components (IgA heavy & light chain, J chain and secretory component), SIgA production is a complex multi-step process which has been attempted in different protein production systems.

Mammalian cells, specifically CHO (Chinese hamster ovary cells), are the industry standard in therapeutic IgG monoclonal antibody production. SIgA manufacture has been achieved either through multiple gene transfection or in vitro reconstitution (ie. dimeric IgA incubated with secretory component) [22]. However, the technology is associated with very high production costs which is reflected in the expensive price of monoclonal antibody therapy, even for IgGs [31]. This problem is exacerbated for SIgA as the yields in mammalian cells are still low.

Plants have emerged as an attractive alternative platform for SIgA production. Plant-derived therapeutics are coming of age—previously against Ebola, Gaucher's disease and more recently the plant-based SARS-CoV-2 vaccine Covifenz® has been approved for use by Health Canada [32–35]. Plants are well suited to produce SIgA by expressing the four components either transiently or by the sequential crossing of plant lines stably expressing each component [36]. A plant-specific issue, however, is the apparent cleavage of the IgA Fc tailpiece required for J chain incorporation—which may be due to differential glycosylation in plants [37]. Efforts to circumvent this include the co-expression of the N-glycosylation facilitating enzyme oligosaccharyltransferase from *Leishmania major* [38].

Furthermore, plant-associated glycans (non-human modifications) on recombinant antibodies present a challenge to therapeutic advancement. For example, plants lack branched and sialylated N-glycosylation, and produce plant-specific xylose and fucose residues [39]. Plant O-glycosylation also differs from humans with the presence of complex arabinogalactans on hydroxyproline residues (extensively found on the IgA1 long hinge [40]) and galactosylated serine which does not appear in humans [41]. This is being addressed by using glycoengineered plant lines—these can produce antibodies which function as well as a their CHO-produced counterpart, but with more controlled and homogenous glycan profiles [42, 43].

Overall, SIgA production in different production platforms carry specific advantages and disadvantages—the overarching issue being the production of multiple assembly intermediates (single chains, monomers and dimers), and it is troublesome to isolate the desired fully assembled SIgA complex [16, 44]. In order to increase the therapeutic potential of SIgA, it is necessary to increase yield (expression levels) and optimise downstream processing.

4. Purification

Downstream purification is the major determinant of cost of goods and monoclonal antibody therapies are currently regarded as some of the most expensive drugs in the world [45]. For SIgA to enter the immunotherapy market, the cost of downstream processing must be comparable to that of IgG. However, there are hurdles associated with SIgA purification which must be addressed beforehand.

Firstly, there is no established gold standard purification resin for SIgA or IgA antibodies. Protein A (the well-established resin used for industry scale IgG purification) cannot be used for SIgA due to the lack of suitable binding sites on the antibody complex. Alternative purification resins that can be used for circumventing this are jacalin [46], SSL7 [47] or generic anti-kappa affinity resins such as protein L [48]; each method is associated with different limitations.

Jacalin is an O-linked glycan-specific lectin derived from jackfruit which is used to purify monoclonal IgA1, and can demonstrate purification efficiency similar to IgG and protein A [49]. Specifically binding to α-D-galactose on the antibody surface, jacalin will not bind to IgA2 or IgA1 with modified glycosylation (e.g. without the O-glycosylated hinge), which may be a benefit or a hindrance depending on the desired product [49]. Inconveniently, Jacalin will also bind to host cell proteins exhibiting O-linked glycosylation, which presents a problem if the target antibody is not present at high concentrations [46].

Protein L is a bacterial cell wall molecule with high binding affinity to certain kappa light chain sequences [48]. This can purify any IgA subtype provided the target sequences are present in the light chain. Unfortunately, protein L will also bind SIgA assembly intermediates and fragments, single light chains, along with monomers and dimeric IgA. This complicates downstream processing since the eluted sample must be further processed (by size exclusion chromatography, for example) to isolate fully formed SIgA. Other purification options, such as CaptureSelect® using llama antibody Fab fragments against single alpha or kappa chain, also suffer from this limitation [50].

Antibody engineering efforts directed at facilitating IgA purification have highlighted the sequences required for affinity purification binding. For example, engineering SIgA light chains to gain protein L binding ability has been described [48]. In addition, murine IgA was made purifiable by SSL7 (another IgA binding protein derived from bacteria) using a 2-amino acid mutation in the Fc region [51], showing that engineering novel purification methods of IgAs are feasible and potentially able to simplify downstream processing.

5. Glycoengineering: finding the sweet spot with complex sugars

Glycosylation is a post-translational protein modification which involves the incorporation of complex sugar molecules (N-linked or O-linked glycans) to specific amino acid residues. It is a common but complex process due to the potential heterogeneity of glycan composition and site occupancy [52]. Antibody glycosylation is dependent on the expression system, as post-translation modifications differ between cell types and production platforms [41]. Glycoengineering approaches in plants for immune modulation usually focuses on changes in galactosylation, fucosylation and sialylation [53]. The glycans associated with an antibody can significantly impact functional characteristics and is a parameter which must be closely monitored in potential therapeutic antibodies [54].

The singular N-linked Fc glycosylation site in IgG has been extensively studied and engineered for a tailored immune response. For example, removing the core fucose increases antibody-dependent cell-mediated cytotoxicity (ADCC) by facilitating the antibody's binding to FcγRIIIA [55]. Conversely, the inflammatory response mediated by IgG effector functions can be reduced by using sialylated glycans, as this impairs complement-dependent cytotoxicity [55]. It is clear even a single glycan site is influential to IgG function. In comparison, a fully assembled SIgA complex can harbour as many as 26 sites.

Every component of SIgA has potential N-linked glycosylation sites: 4 to 9 per IgA monomer, 1 on the joining chain and 7 on the secretory component. In addition, the IgA1 subtype has an extended hinge region with up to 6 clustered O-glycan sites which make investigations into glycosylation status and the resulting effect on function difficult [56]. Despite this, it is clear that glycosylation is important for effective SIgA function. For instance, SIgA with unglycosylated secretory component is subject to rapid breakdown in the gut [57]. Indeed, the glycan repertoire of IgA is markedly different to that of the well-established IgG and requires extensive further study for levels of familiarity to be comparable [58].

6. Applications: engineering SIgA for passive immunisation?

The route of administration of monoclonal antibody therapy is a determining factor in its effectiveness. Vaccination via the systemic route, whilst effective in generating serum responses, often does not mount a sufficient mucosal response [59]. As a result, most systemic vaccines will not protect against initial infection, since host-pathogen interactions mainly occur directly via the mucosa. This highlights a need for mucosal immunisation and by extension passive immunisation with antibodies.

Passive immunisation would involve the topical administration of pathogen-specific SIgA directly to the oral, nasal, respiratory or gastro-intestinal mucosal. This would immediately form an enhanced protective barrier. For example, there has been significant efforts in the search for artificial colostrum—an oral formulation containing mucosal antibodies against disease in neonates, emulating the protective qualities of breastmilk [60].

In animals, orally-administered VHH-SIgA fusions have provided protection against *E.coli* infection superior to their IgG counterparts [61]. This proof-of-concept study shows the exciting applications of passive immunisation and highlights the importance of IgA-based therapy, although it is yet to be replicated in humans. Furthermore, SIgA's heterologous glycan profile is less of a complication if taken via the oral route due to the body's natural oral tolerance to exogenous antigens [62]. However, the pharmacokinetics and stability of SIgA in the human gastric environment may need further consideration and optimization before implementation [63]. Although our understanding of IgA-mucin interactions is still limited, it is a promising field of study nonetheless.

7. Summary

The mucosal system is the body's first line of defence against infection and SIgA is the primary immunological weapon. Although there have been significant advancements in the understanding of this antibody complex, SIgA has been overshadowed by the therapeutic success of IgG antibodies. Renewed interest in mucosal vaccines and oral immunotherapy has highlighted that existing therapeutics may not be best suited for mucosal therapy, ushering a potential new era of SIgA-based therapy. This chapter has highlighted that:

- Fine-tuning the correct balance of antibody effector function is critical for therapeutic viability. Further engineering of SIgA's structure, glycans and production will continue to push it closer to joining IgG in the mainstream immunotherapy market.

- Novel engineering of IgA-based therapeutics are attractive due to new intellectual property opportunities in the rapidly growing and lucrative field of immunotherapy.

- As machine learning applications to antibody engineering become more accessible and continues to inform the design on future antibodies, SIgA will surely benefit. However, more real-world experimental data may be required for accurate computational design of SIgA in particular.

Acknowledgements

We would like to thank to the generous support of European Research Council's Horizon 2020 programme under grant agreement 760331 (Newcotiana) and 774078 (Pharma Factory), as well as the Sir Joseph Hotung Charitable Settlement. We also would like to thank colleagues in the Hotung Molecular Immunology Unit at St. George's, University of London for their support and constructive discussion.

Author details

Nicole Falci Finardi, Audrey Y-H. Teh*† and Julian K-C. Ma†
St George's, University of London, UK

*Address all correspondence to: ateh@sgul.ac.uk

† These authors have contributed equally to this work and share senior authorship.

References

[1] Russell MW. Biological functions of IgA. In: Mucosal Immune Defense: Immunoglobulin A. Boston, MA: Springer; 2007. pp. 144-172

[2] Wang Y, Wang G, Li Y, Zhu Q, Shen H, Gao N, et al. Structural insights into secretory immunoglobulin A and its interaction with a pneumococcal adhesin. Cell Research. 2020;**30**(7):602-609

[3] de Sousa-Pereira P, Woof JM. IgA: Structure, function, and developability. Antibodies. 2019;**8**(4):57

[4] Renegar KB, Small PA, Boykins LG, Wright PF. Role of IgA versus IgG in the control of influenza viral infection in the murine respiratory tract. The Journal of Immunology. 2004;**173**(3):1978-1986

[5] de Fays C, Carlier FM, Gohy S, Pilette C. Secretory immunoglobulin A immunity in chronic obstructive respiratory diseases. Cell. 2022;**11**(8):1324

[6] Rogier EW, Frantz AL, Bruno ME, Kaetzel CS. Secretory IgA is concentrated in the outer layer of colonic mucus along with gut bacteria. Pathogens. 2014;**3**(2):390-403

[7] Ejemel M, Li Q, Hou S, Schiller ZA, Tree JA, Wallace A, et al. A cross-reactive human IgA monoclonal antibody blocks SARS-CoV-2 spike-ACE2 interaction. Nature Communications. 2020;**11**(1):1-9

[8] Corthesy B. Role of secretory IgA in infection and maintenance of homeostasis. Autoimmunity Reviews. 2013;**12**(6):661-665

[9] Vecchio AL, Conelli ML, Guarino A. Infections and chronic diarrhea in children. The Pediatric Infectious Disease Journal. 2021;**40**(7): e255-e258

[10] Farthing M, Salam MA, Lindberg G, Dite P, Khalif I, Salazar-Lindo E, et al. Acute diarrhea in adults and children: A global perspective. Journal of Clinical Gastroenterology. 2013;**47**(1):12-20

[11] Stoppato M, Gaspar C, Regeimbal J, Nunez RG, Giuntini S, Schiller ZA, et al. Oral administration of an anti-CfaE secretory IgA antibody protects against Enterotoxigenic Escherichia coli diarrheal disease in a nonhuman primate model. Vaccine. 2020;**38**(10):2333-2339

[12] Bhaskara V, Leal MT, Seigner J, Friedrich T, Kreidl E, Gadermaier E, et al. Efficient production of recombinant secretory IgA against Clostridium difficile toxins in CHO-K1 cells. Journal of Biotechnology. 2021;**331**:1-3

[13] Wallace AL, Schneider MI, Toomey JR, Schneider RM, Klempner MS, Wang Y, et al. IgA as a potential candidate for enteric monoclonal antibody therapeutics with improved gastrointestinal stability. Vaccine. 2020;**38**(47):7490-7497

[14] Su CT, Lua WH, Ling WL, Gan SK. Allosteric effects between the antibody constant and variable regions: A study of IgA Fc mutations on antigen binding. Antibodies. 2018;**7**(2):20

[15] Senior BW, Woof JM. Sites in the CH3 domain of human IgA1 that influence sensitivity to bacterial IgA1 proteases. The Journal of Immunology. 2006;**177**(6):3913-3919

[16] Senior BW, Dunlop JI, Batten MR, Kilian M, Woof JM. Cleavage of a

recombinant human immunoglobulin A2 (IgA2)-IgA1 hybrid antibody by certain bacterial IgA1 proteases. Infection and Immunity. 2000;**68**(2):463-469

[17] Lohse S, Brunke C, Derer S, Peipp M, Boross P, Kellner C, et al. Characterization of a mutated IgA2 antibody of the m (1) allotype against the epidermal growth factor receptor for the recruitment of monocytes and macrophages. Journal of Biological Chemistry. 2012;**287**(30):25139-25150

[18] Westerhof LB, Wilbers RH, van Raaij DR, van Wijk CZ, Goverse A, Bakker J, et al. Transient expression of secretory IgA in planta is optimal using a multi-gene vector and may be further enhanced by improving joining chain incorporation. Frontiers in Plant Science. 2016;**6**:1200

[19] Redwan ER, Matar SM, Serour IA. Recombinant human J-chain: Fix the protein aggregations and yield maximize. Human Antibodies. 2006;**15**(3):95-102

[20] Juarez P, Huet-Trujillo E, Sarrion-Perdigones A, Falconi EE, Granell A, Orzaez D. Combinatorial analysis of secretory immunoglobulin A (sIgA) expression in plants. International Journal of Molecular Sciences. 2013;**14**(3):6205-6222

[21] Wei H, Wang JY. Role of polymeric immunoglobulin receptor in IgA and IgM transcytosis. International Journal of Molecular Sciences. 2021;**22**(5):2284

[22] Virdi V, Juarez P, Boudolf V, Depicker A. Recombinant IgA production for mucosal passive immunization, advancing beyond the hurdles. Cellular and Molecular Life Sciences. 2016;**73**(3):535-545

[23] Hoey RJ, Eom H, Horn JR. Structure and development of single domain antibodies as modules for therapeutics and diagnostics. Experimental Biology and Medicine. 2019;**244**(17):1568-1576

[24] De Greve H, Virdi V, Bakshi S, Depicker A. Simplified monomeric VHH-Fc antibodies provide new opportunities for passive immunization. Current Opinion in Biotechnology. 2020;**61**:96-101

[25] Kelton W, Mehta N, Charab W, Lee J, Lee CH, Kojima T, et al. IgGA: A “cross-isotype” engineered human Fc antibody domain that displays both IgG-like and IgA-like effector functions. Chemistry & Biology. 2014;**21**(12):1603-1609

[26] Jumper J, Evans R, Pritzel A, Green T, Figurnov M, Ronneberger O, et al. Highly accurate protein structure prediction with AlphaFold. Nature. 2021;**596**(7873):583-589

[27] Warszawski S, Borenstein Katz A, Lipsh R, Khmelnitsky L, Ben Nissan G, Javitt G, et al. Optimizing antibody affinity and stability by the automated design of the variable light-heavy chain interfaces. PLoS Computational Biology. 2019;**15**(8):e1007207

[28] Gao W, Mahajan SP, Sulam J, Gray JJ. Deep learning in protein structural modeling and design. Patterns. 2020;**1**(9):100142

[29] Akbar R, Robert PA, Weber CR, Widrich M, Frank R, Pavlović M, et al. In silico proof of principle of machine learning-based antibody design at unconstrained scale. MAbs. 2022;**14**(1):2031482

[30] Hanning KR, Minot M, Warrender AK, Kelton W, Reddy ST. Deep mutational scanning for therapeutic antibody engineering. Trends in Pharmacological Sciences. 2022;**43**(2):123-135

[31] Hernandez I, Bott SW, Patel AS, Wolf CG, Hospodar AR, Sampathkumar S, et al. Pricing of monoclonal antibody therapies: Higher if used for cancer. The American Journal of Managed Care. 2018;**24**(2):109-112

[32] Group TP, Multi-National PREVAIL II Study Team. A randomized, controlled trial of ZMapp for Ebola virus infection. The New England Journal of Medicine. 2016;**375**(15):1448

[33] Fox JL. First plant-made biologic approved. Nature Biotechnology. 2012;**30**(6):472-473

[34] Ibtisam T. Effectiveness of currently authorized vaccines against SARS-CoV-2 variants of concern and the future of COVID-19 vaccines. Undergraduate Journal of Experimental Microbiology and Immunology. 2022;**6**:1

[35] Ward BJ, Gobeil P, Séguin A, Atkins J, Boulay I, Charbonneau PY, et al. Phase 1 randomized trial of a plant-derived virus-like particle vaccine for COVID-19. Nature Medicine. 2021;**27**(6):1071-1078

[36] Paul M, Reljic R, Klein K, Drake PM, Van Dolleweerd C, Pabst M, et al. Characterization of a plant-produced recombinant human secretory IgA with broad neutralizing activity against HIV. MAbs. 2014;**6**(6):1585-1597

[37] Dicker M, Tschofen M, Maresch D, König J, Juarez P, Orzaez D, et al. Transient glyco-engineering to produce recombinant IgA1 with defined N-and O-glycans in plants. Frontiers in Plant Science. 2016;**7**:18

[38] Göritzer K, Goet I, Duric S, Maresch D, Altmann F, Obinger C, et al. Efficient N-glycosylation of the heavy chain tailpiece promotes the formation of plant-produced dimeric IgA. Frontiers in Chemistry. 2020;**8**:346

[39] Mercx S, Smargiasso N, Chaumont F, De Pauw E, Boutry M, Navarre C. Inactivation of the β (1, 2)-xylosyltransferase and the α (1, 3)-fucosyltransferase genes in Nicotiana tabacum BY-2 cells by a multiplex CRISPR/Cas9 strategy results in glycoproteins without plant-specific glycans. Frontiers in Plant Science. 2017;**8**:403

[40] Karnoup AS, Turkelson V, Anderson WK. O-linked glycosylation in maize-expressed human IgA1. Glycobiology. 2005;**15**(10):965-981

[41] Strasser R. Plant protein glycosylation. Glycobiology. 2016;**26**(9):926-939

[42] Montero-Morales L, Steinkellner H. Advanced plant-based glycan engineering. Frontiers in Bioengineering and Biotechnology. 2018;**6**:81

[43] Teh AY, Cavacini L, Hu Y, Kumru OS, Xiong J, Bolick DT, et al. Investigation of a monoclonal antibody against enterotoxigenic Escherichia coli, expressed as secretory IgA1 and IgA2 in plants. Gut Microbes. 2021;**13**(1):1859813

[44] Chintalacharuvu KR, Gurbaxani B, Morrison SL. Incomplete assembly of IgA2m (2) in Chinese hamster ovary cells. Molecular Immunology. 2007;**44**(13):3445-3452

[45] Samaranayake H, Wirth T, Schenkwein D, Räty JK, Ylä-Herttuala S. Challenges in monoclonal antibody-based therapies. Annals of Medicine. 2009;**41**(5):322-331

[46] Puligedda RD, Vigdorovich V, Kouiavskaia D, Kattala CD, Zhao JY, Al-Saleem FH, et al. Human IgA monoclonal antibodies that neutralize poliovirus, produced by Hybridomas and

recombinant expression. Antibodies. 2020;**9**(1):5

[47] Langley R, Wines B, Willoughby N, Basu I, Proft T, Fraser JD. The staphylococcal superantigen-like protein 7 binds IgA and complement C5 and inhibits IgA-FcαRI binding and serum killing of bacteria. The Journal of Immunology. 2005;**174**(5):2926-2933

[48] Åkerström B, Björck L. Protein L: An immunoglobulin light chain-binding bacterial protein: Characterization of binding and physicochemical properties. Journal of Biological Chemistry. 1989;**264**(33):19740-19746

[49] Roque-Barreira MC, Campos-Neto AN. Jacalin: An IgA-binding lectin. The Journal of Immunology. 1985;**134**(3):1740-1743

[50] Hermans P, Adams H, Detmers F. Purification of antibodies and antibody fragments using CaptureSelect™ affinity resins. In: Monoclonal Antibodies. Totowa, NJ: Humana Press; 2014. pp. 297-314

[51] Bakshi S, Depicker A, Schepens B, Saelens X, Juarez P. A two-amino acid mutation in murine IgA enables downstream processing and purification on staphylococcal superantigen-like protein 7. Journal of Biotechnology. 2019;**294**:26-29

[52] Ferrara C, Grau S, Jäger C, Sondermann P, Brünker P, Waldhauer I, et al. Unique carbohydrate–carbohydrate interactions are required for high affinity binding between FcγRIII and antibodies lacking core fucose. Proceedings of the National Academy of Sciences. 2011;**108**(31):12669-12674

[53] Göritzer K, Strasser R. Glycosylation of plant-produced immunoglobulins. In: Antibody Glycosylation. Cham: Springer; 2021. pp. 519-543

[54] Stelter S, Paul MJ, Teh AY, Grandits M, Altmann F, Vanier J, et al. Engineering the interactions between a plant-produced HIV antibody and human Fc receptors. Plant Biotechnology Journal. 2020;**18**(2):402-414

[55] Quast I, Keller CW, Maurer MA, Giddens JP, Tackenberg B, Wang LX, et al. Sialylation of IgG Fc domain impairs complement-dependent cytotoxicity. The Journal of Clinical Investigation. 2015;**125**(11):4160-4170

[56] de Haan N, Falck D, Wuhrer M. Monitoring of immunoglobulin N-and O-glycosylation in health and disease. Glycobiology. 2020;**30**(4):226-240

[57] Pabst O, Slack E. IgA and the intestinal microbiota: The importance of being specific. Mucosal Immunology. 2020;**13**(1):12-21

[58] Chandler KB, Mehta N, Leon DR, Suscovich TJ, Alter G, Costello CE. Multi-isotype Glycoproteomic characterization of serum antibody heavy chains reveals isotype-and subclass-specific N-glycosylation profiles*[S]. Molecular & Cellular Proteomics. 2019;**18**(4):686-703

[59] Kurono Y. The mucosal immune system of the upper respiratory tract and recent progress in mucosal vaccines. Auris, Nasus, Larynx. 2022;**49**(1):1-0

[60] Richards A, Baranova D, Mantis NJ. The prospect of orally administered monoclonal secretory IgA (SIgA) antibodies to prevent enteric bacterial infections. Human Vaccines & Immunotherapeutics. 2021;**18**:1-8

[61] Virdi V, Coddens A, De Buck S, Millet S, Goddeeris BM, Cox E, et al. Orally fed seeds producing designer IgAs protect weaned piglets against enterotoxigenic Escherichia coli infection. Proceedings of the

National Academy of Sciences. 2013;**110**(29):11809-11814

[62] Juarez P, Virdi V, Depicker A, Orzaez D. Biomanufacturing of protective antibodies and other therapeutics in edible plant tissues for oral applications. Plant Biotechnology Journal. 2016;**14**(9):1791-1799

[63] Hu Y, Kumru OS, Xiong J, Antunez LR, Hickey J, Wang Y, et al. Preformulation characterization and stability assessments of secretory IgA monoclonal antibodies as potential candidates for passive immunization by oral administration. Journal of Pharmaceutical Sciences. 2020;**109**(1):407-421

Chapter 9

Perspective Chapter: Phytocompounds as Immunomodulators

Ayda Cherian and Velmurugan Vadivel

Abstract

Healthy operation of every organ depends on immune cells. T-cells, B-cells, and natural killer cells that control the immune homeostasis. Immunotherapy includes the process by which immune cells are immunomodulated. Immunological responses can be induced by immunostimulants, amplified by immune boosters, attenuated by immunomodulators, and prevented by immunosuppressive agents, according to therapeutic techniques. The over-activation of the immune system is mostly to blame for the rise of chronic immunological illnesses such as viral infections, allergies, and cancer. Immunomodulators may also be used to control the severity of long-term immunological diseases. Additionally, it is discovered that these immunomodulator-acting proteins represent prospective molecular targets for the control of the immune system. Furthermore, it is well known that organic molecules like phytocompounds have the ability to bind to these locations and affect the immune system. Curcumin, quercetin, stilbenes, flavonoids, and lignans are examples of specific phytocompounds shown to have immunomodulatory properties to address immunological diseases.

Keywords: autoimmune diseases, herbal compounds, immune system, immunotherapy, phytomedicines

1. Introduction

A healthy immune system is beneficial for every organ. The self-regulation of T-cell, B-cell, and natural killer cell activity creates the immune system's homeostasis. The functioning immune systems serve as a connected network in the body, defending the organs against numerous immunological diseases. Additionally, it recognises invasive pathogens like bacteria, viruses, and parasites and reacts quickly to them [1]. The innate immune system and the adaptive immune system are the immune system's two main subsystems. The innate immune system responds in a programmed way to a wide range of immunologically active proteins and immune stimuli. In comparison, the adaptive immune system responds sequentially to each stimulus by identifying the chemicals the immune cells use to operate [2]. Immunological disorders, inflammatory reactions, and the development of cancer can all be brought on by the dysregulation of immune activities. Chronic immunodeficiency has also been linked to an increase in infections that can be fatal [3]. Genetic changes are also related to some immune illnesses. These severe immunological conditions include acquired immunodeficiency syndrome (AIDS) [4]. Type 1 diabetes, rheumatoid arthritis, Hashimoto's thyroiditis and systemic lupus erythematosus are just a few major autoimmune disorders that are commonly treated using immunomodulatory medications [5].

However, the excellent health of a patient with chronic immunological problems depends on a functioning immune system. Numerous organic substances, including polyphenols, alkaloids, polysaccharides, glycosides, lactones, terpenoids, and flavonoids, have been shown to modify immune cells and have immunomodulatory effects [6]. The essential immunological cells it interacts with are macrophages, dendritic cells, lymphocytes (T- and B-cells), and natural killer (NK) cells. The primary immunomodulatory effects are brought about by secreted antibodies against various pathogens and maintain immunological homeostasis [7]. Our bodies' immune systems create a self-defence strategy to stave off illnesses from numerous pathogens. Innate and adaptive immune cells work with anatomical and physiological barriers to form the human body's three layers of protection against pathogens. Saliva, epidermis, mucous membranes, low stomach pH, and other secretions are among the anatomical and physiological barriers that make up the immune system's first line of defence against pathogens. Other bodily fluid discharges and the stomach's low pH level can also easily interact with immune cells [8]. The second level of defence begins with inflammatory cells like mast and macrophage cells and uses complement system formation and immunomodulatory action to establish innate immunity. It covers the roles of basophils, NK cells, mast cells, neutrophils, eosinophils, macrophages, dendritic cells, and natural killer T cells in immunological activity [9]. Twenty serum glycoproteins make up the complement system, where the main immunomodulatory effects begin. Additionally, three pathways—the lectin pathway (mannose-binding lectin reaction), the alternative pathway (bacterial endotoxin reaction), and the classical pathway (antigen–antibody reactions)—are involved in the activation of the complement system. Circulatory complement-3 (C3) proteins are involved in every pathway and are crucial for immunomodulatory effects [10]. Through T and B cells, the first and second levels of defence systems generate non-specific immune responses, known as adaptive immunity. This is an antigen–antibody-associated immune cell-specific response and forms the third level of the immune defence system [9]. The primary innate and adaptive immune system function and their cells are illustrated in **Figure 1**.

Pathogen recognition receptors (PRRs) are used in the innate immune system to identify the infection pattern. Pathogen-associated molecular patterns (PAMPs) detect the microbial components of molecular proteins. The PAMPs consists of bacterial parts such flagella, nucleic acid components, and lipopolysaccharide [7]. PPR families include cytoplasmic proteins like NOD-like receptors (NLRs), retinoic acid-inducible gene-I-like receptors (RLRs), and transmembrane proteins like C-type lectin receptors (CLRs), and toll-like receptors (TLRs) [11]. Antigen-presenting cells, such as dendritic cells (DCs), B cells, and macrophages, make up non-specific immune systems. In order for cytotoxic T-cells and B-cells, as well as non-antigen specific macrophages, NK cells, and eosinophils, to perform their essential roles, dendritic cells must first deliver the antigens to a group of domain-4 (CD4)-T-helper cells. Tumour necrosis factor-α (TNF-α) and cytokines (IFN-γ) are generated during the proliferation of DCs as a result of immunological responses [12]. In adaptive immunity, a defence mechanism is formed by the injected agents against the particular pathogens and stop infections. It is also referred to as acquired immunity. It

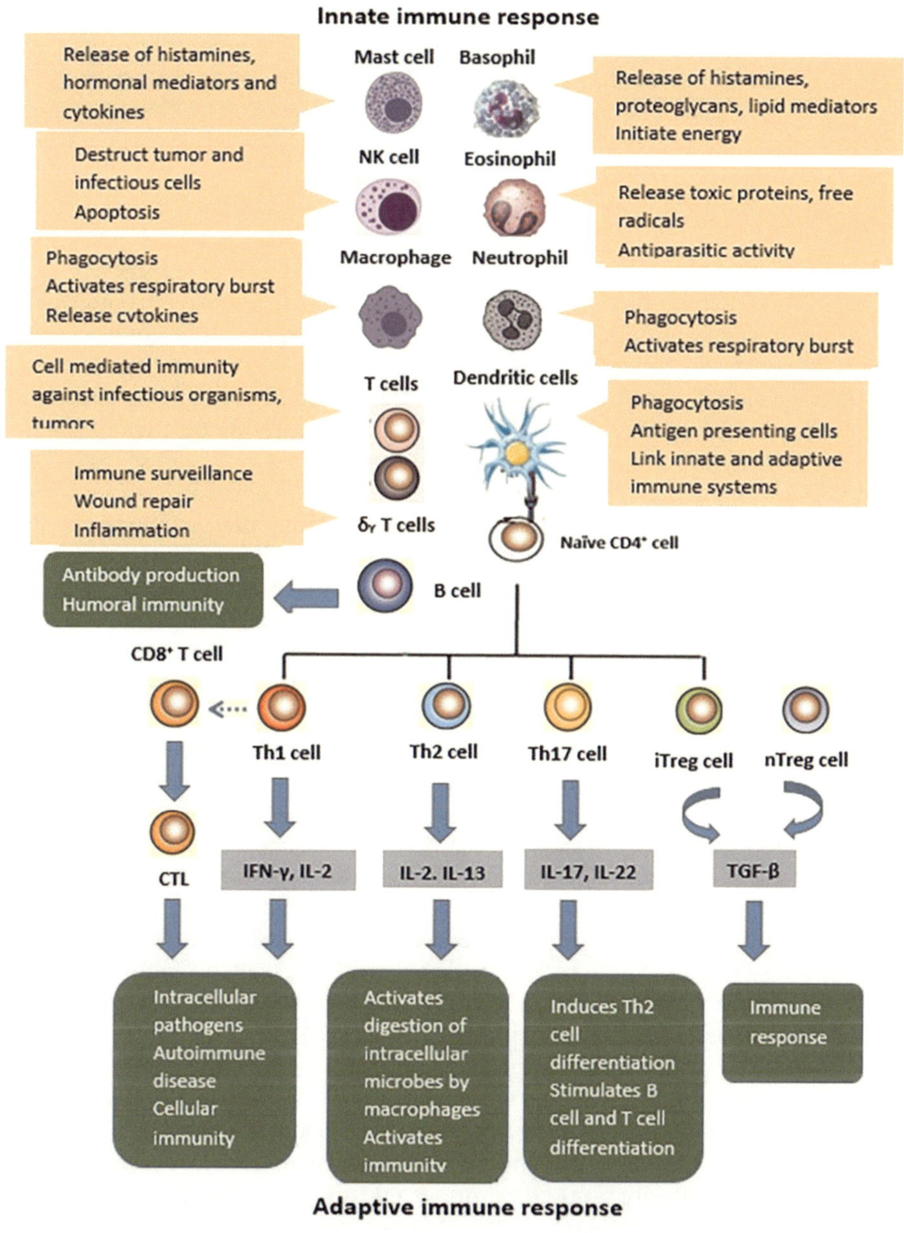

Figure 1.
Immune system traits and functions of different innate and adaptive immune cells. There are two types of immunity in the immune system: Innate immunity and adaptive immunity. Among the participants in the innate immune system are dendritic cells, macrophages, mast cells, granulocytes (neutrophils, eosinophils and basophils), NK cells, NKT cells and δγ T cells. The pink grid lists each cell type's primary fuctions. B cells, CD4⁺ T cells, CD8⁺ T cells and other cells are all part of the adaptive immune system. Under various microenvironments specialised by interacting cytokines and chemokines, as well as unique activation of particular transcription factors, CD4⁺ T cells can develop into Th1, Th2, Th17 and inducible Treg (iTreg) cells. CD8⁺ T cells are in charge of verifying cytotoxicity against cancerous or virus-infected cells. Treg cells are often divided into two types, natural Treg (nTreg) cells and iTreg cells. To maintain immunological homeostasis, they can control particular immune responses, including immune tolerance. Abbreviations: TCR, T cell receptor; MHC-II, major histocompatibility complex class II; NK, natural killer cells; NK T cells, natural killer T cells; Th1, T-helper type 1; Th2, T-helper type 2; Th17, T-helper type 17; IL-2, interleukin-2; IL-13, interleukin-13; IL-17, interleukin-17; IL-22, interleukin-22.

primarily affects T and B cells, resulting in a cell-mediated immunological response and antibody generation. This adaptive immune response is primarily supported by CD4+ T-lymphocytes (helper T-cells) and CD8+ T-lymphocytes. Additionally, these T-helper cells are developed from Type-1 T helper (Th1) and Th2 cells, and they secrete IFN-γ and IL-5 to improve adaptive immune responses [13].

1.1 Immunomodulators

Immunomodulators include immunostimulants and immunosuppressive medications [14]. The efficient and effective homeostasis that the healthy immune system generates keeps the human body free from disease. Additionally, it governs the cellular signalling molecules as the favourable host responses and has strong communication of cells via signal transduction pathways. Immune cell-acting substances can either boost or inhibit immune cells' typical efficiency and function when administered endogenously or exogenously. It is sometimes referred to as host responses for immunomodulatory actions [15]. This immunity prevents chronic immunological disorders such as AIDS, cancer, autoimmune diseases, and allergic reactions [14].

Immunostimulants, immunoadjuvants, and immunosuppressants are the three categories that comes under immunomodulators. Immunostimulants are substances that cause the immune system's cells to become active. Vaccines operate according to these principles to enhance the specific immunostimulant actions. Immune response to specific pathogenic antigens is strengthened by it. Some reports have shown that natural substances, such as phytocompounds, are known to have general immunostimulant effects. Cancer and other chronic infections like those that cause immunodeficiency disorders are also said to be attenuated. Interferon-alpha (IFN-alpha) and granulocyte colony-stimulating factors are two examples of endogenous immunostimulants involved in developing immunostimulant effects [16]. Moreover, FDA also describes the immunoadjuvants therapy category for immunological illnesses. The conjugation of immunoadjuvants with a vaccination antigen results in the augmentation and potentiation of target proteins for the particular immune response against the antigen. Histamine, tuftsin, interferons, IL-1, transfer factor, and IL-1 are endogenous natural adjuvants. It can intensify the targeted antigen's interactions with the host immune system and induce phagocytosis [17, 18].

To lessen overly strong immunological reactions, immunosuppressants are also necessary. Excessive immune cell function can cause serious systemic consequences. Immunosuppressive medications are necessary for various therapeutic situations, such as organ transplantation (pre and post-surgical conditions). A few immunosuppressive drugs momentarily weaken immune cell functions. Along with these conditions, it is also used to treat rheumatoid arthritis, myasthenia gravis, and Grave's disease. Additionally, it manages graft rejection reactions in tissue (skin) and cells (bone marrow transplant) [17]. Similarly, some natural phytoconstituents suppress the immune system and treat immunological diseases by interfering with the host cell's molecular communication pathways.

2. Phytocompounds as immunomodulators

More than 5000 years of history are found for medications made from natural products, compared to a few hundred years for Western medicine. More than 85,000 plant species have been used medicinally around the world. According to the WHO, up to 80% of people worldwide, primarily in developing nations, rely on herbal remedies to cure a various illnesses, including immunological disorders [19]. Additionally, about 30% of all FDA-approved medications have a botanical origin [20]. Based on this data, it's critical to look into traditional phytomedicines' chemical makeup to assess their potential as immunomodulatory agents for immunological diseases. The chemical compositions, molecular targets, and related illnesses of the representative phytocompounds are summarised. The classification of phytochemicals is given in **Figure 2**.

Numerous pharmacological effects can be attributed to phytocompounds including immunomodulatory effects. Several phytoconstituents, including polyphenols like stilbenes, resveratrol, hydroxycinnamic acids, and curcumin; flavonoids like epigallocatechin gallate (EGCG) and quercetin; alkaloids like berberine (BBR), & colchicine; terpenoids like andrographolide, & oleanolic acid; polysaccharides like pectin [21, 22] are found to act on immune system. They modulate the activity of a variety of immune cells, including dendritic cells (DCs), lymphocytes, neutrophils, monocytes, macrophages, basophils, mast cells, eosinophils, and natural killer (NK) cells. It frequently controls phagocytic cells, including neutrophils, monocytes, macrophages, basophils, mast cells and eosinophils that secrete inflammatory mediators and natural killer (NK) cells [23].

It is evidenced experimentally that the phytocompounds alter a number of the molecular targets of the immune cell signalling mechanism. In immune cells, they also modify the release of soluble substances and these include transcription factors and interleukins (IL) such as IL-2, IL-4, IL-6, IL-12, IL-17 and immunoglobulins (Igs) [24]. The transcription factors nuclear factor-κB (NF-κB) and inhibitor κB (I-κB) are frequently used to control the activity of immune cells and have immunomodulatory effects. It also phosphorylates the c-Jun N-terminal kinases (JNK/Jun) pathways, degrades the inhibition of the NF-κB p65 subunit and I-κB, increases the levels of reduced glutathione (GSH) and superoxide dismutase (SOD) dependent on T-lymphocytes, and collaterally enhances immune responses through the expression

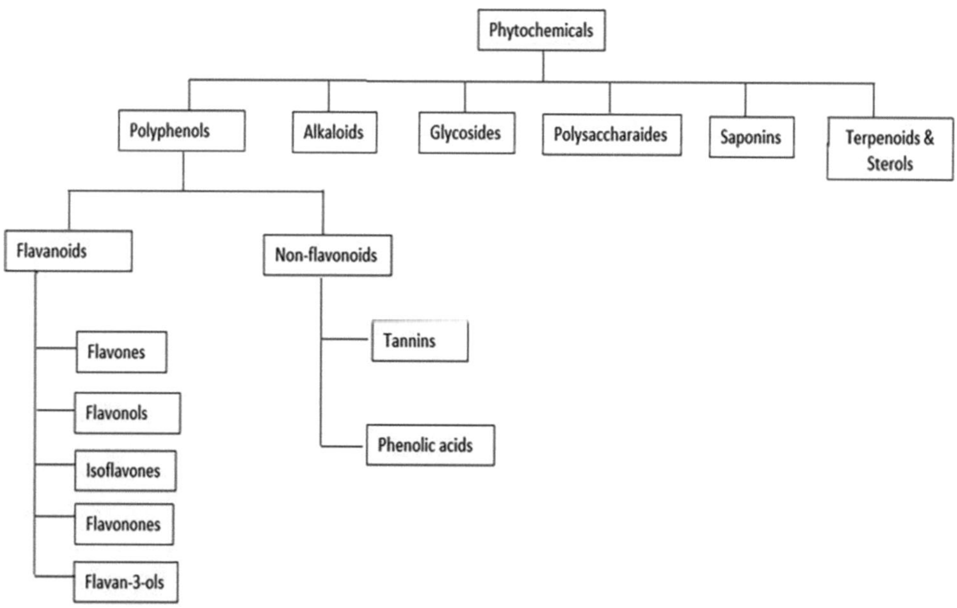

Figure 2.
Classification of phytochemicals.

of TLR4 and upregulation of cytokine genes [25]. Additionally, different immune cells and their cell signals behave in varied ways based on the pathophysiological circumstances of various body systems. Therefore, a clear, precise mechanism and the unique activity of phytoconstituents on immune cells must be thoroughly studied [26]. The following sections have detailed the specifics of phytoconstituents for the immunomodulatory mechanism of action.

2.1 Polyphenol

Any compound, including functional derivatives (esters, glycosides, etc.), with an aromatic ring and one or more hydroxyl substituents is referred to as a "polyphenol" or "phenolic". Polyphenols in foods or natural health products come from one of the main classes of secondary plant metabolites derived from tyrosine or phenylalanine. They are widely found in fruits, vegetables, beverages and cereals [27].

2.1.1 Stilbene derivatives

The phenolic compounds known as stilbenes have two aromatic rings connected by an ethene bridge (C6–C2–C6) [28]. Trans-3,5,4′-trihydroxystilbene, also known as resveratrol, is a well-known phytoalexin of the stilbene class. It is a natural substance found in grapes, berries, and other traditional Chinese medicines like Polygonum cuspidatum. It is known to work by modulating a various distinct pathways to produce its effects [29]. Numerous cell-signalling molecules connected to inflammation have been demonstrated to bind to resveratrol. It inhibits cyclooxygenase-2 (COX-2) protein expression in mammary epithelial cells and activates phorbol-12-myristate-13-acetate (PMA) to control the protein kinase C (PKC) transduction pathway for immunomodulatory effects [30]. Additionally, it inhibits p65 subunit phosphorylation, IκBα kinase phosphorylation, and NF-κB for DNA activities, all of which are implicated in immunological actions [31, 32]. Moreover, it blocks the activator protein-1 (AP-1) that is in charge of immunological effects [33]. Experimental evidence suggests that resveratrol blocks the lipopolysaccharide (LPS) induced NF-κB p65 nuclear translocation, as well as the down-regulation of IL-6, nitric oxide, IL-18, vascular endothelial growth factor (VEGF), and matrix metalloproteinases (MMP-2 and MMP-9) secretion in E11 cells. Furthermore, it inhibits THP-1 and U937 monocyte migration in inflammatory areas by immunological reactions [34]. Resveratrol dramatically decreases the expression and activity of inducible nitric oxide synthase (iNOS) [35].

2.1.2 Hydroxycinnamic acids

Another name for hydroxycinnamic acids is hydroxycinnamates (i.e., curcumin, and p-Coumaric acid). Chemically, it belongs to the phenylpropanoids or class of aromatic acids. It is a hydroxy derivative of cinnamic acid, and has a C6-C3 skeleton. It appears as α-Cyano-4-hydroxycinnamic acid in food, including fruits, vegetables, and cereals, and has the potential to scavenge free radicals and guard against inflammation, dyslipidemia, insulin resistance, diabetes, and cardiovascular illnesses [36]. The molecular function of hydroxycinnamic acids is to improve the host immune response and lessen harm to the body's essential organs [37]. The active component of Curcuma longa is curcumin, a hydroxycinnamic acid derivative and is known as diferuloylmethane chemically [38]. The anti-platelet, anti-inflammatory, hepatoprotective, anti-cancer, and anti-arthritic properties of curcumin make it a popular treatment option in Ayurvedic medicine [39]. It is also recognised to have immunomodulatory effects. Experimental evidence suggests that it inhibits the transcription factors such as cytosine-cytosine-adenosine-adenosine-thymidine (CCAAT)/enhancer-binding protein (C/EBP), CTCF, β-catenin, heat shock factor-1, Notch-1, hypoxia-inducible factor-1 (HIF-1), early growth response-1 (Egr-1), AP-1, signal transducers and activators of transcription (STAT)-1, 3, 4, 5 and NF-κB [40]. Additionally, AP-1 and NF-κB, which are immune cell transcription factors, are suppressed by curcumin to provide the anti-tumour effect [41].

2.1.3 Flavonoids

Flavonoids are one of the most prevalent naturally occurring substances in all vascular plants. At least 6500 naturally occurring flavonoids have been discovered, and nearly all plant tissues are capable of producing flavonoids. The 15 carbon atoms that make up the essential backbone of flavonoids (C6-C3-C6) define them. They are typically divided into seven classes based on their chemical makeup: flavones, flavanones, flavonols, flavanonols, isoflavones, flavanols, and anthocyanidins [42]. They often take the form of a flavonoid glycoside or an aglycone. While aglycones are primarily found in woody tissues, flavonoid glycosides are primarily found in leaves, flowers, or fruits. Both flavonoid aglycones and glycosides can be found in seeds. Flavonoids have long been known to have anti-inflammatory, anti-hepatotoxic, anti-atherogenic, anti-osteoporotic, anti-allergic, and anti-cancer properties, in addition to their well-known antioxidant activity [43]. Quercetin is a flavonol that is present in foods including grapes, tea, onions, apples, and leafy green vegetables [44]. The most well-known active ingredient in tea is epigallocatechin gallate (EGCG), a potent antioxidant. It can affect phase I and phase II enzymes in addition to being a potent anti-inflammatory and antioxidant that guards the body against the damaging effects of free radicals [45]. The inhibition of transcriptional factors (such as NF-kB and AP-1) and the elevation of Nrf-2, which results in a decrease in pro-inflammatory mediators, are thought to be the anti-inflammatory modes of action of quercetin and EGCG [46]. These properties have led researchers to consider using these chemicals to treat inflammatory illnesses, ageing, neurological disorders, inflammatory bowel diseases, cancer, and diabetes.

2.2 Terpenoids

Terpenoids, also known as isoprenoids, comprise five carbon isoprene units (C5H8). It is a broad and diversified family of naturally occurring chemical compounds from the 5-carbon molecule isoprene. The term terpenes are used as another name for the polymers of isoprene. For terpenes, oxygen is the primary functional group. Quite a few pharmacological effects are present [47]. Found in several conventional herbal remedies, including eucalyptus leaves, the flavours of cinnamon, cloves, and sunflowers, as well as foods like ginger and tomatoes. The bioactive components of terpenoids include citral, menthol, camphor, salvinorin A, cannabinoids, ginkgolide, bilobalide, and curcuminoids [48, 49]. In addition, the bioactive immunomodulating drugs andrographolide and oleanolic acid are active. Andrographolide is a bicyclic diterpenoid lactone compound. It is an official Chinese herbal medicine component and has strong anti-inflammatory properties. Additionally, it can be found in the leaves of Andrographis paniculata and is used to treat rheumatoid arthritis, laryngitis, and diarrhoea. Nitric oxide (NO) generation is known to be reduced, and iNOS expression is inhibited by andrographolide in RAW 264.7 cells [50]. According to in-vitro research, andrographolide regulates the activity of immune cells like macrophages and microglia, which produces immunomodulatory effects by reducing the levels of TNF-α, COX-2, IL-12, iNOS, and PGE2 proteins [51]. Additionally, it actively modifies the virality of the influenza virus by downregulating the genes for the JAK/STAT signalling and NF-kB signal pathways [52].

Oleanic acid (3β-hydroxy-olea-12-en-28-oic acid) is another name for oleanolic acid. It is naturally related to betulinic acid and contains a pentacyclic triterpenoid moiety. Olea europaea, Rosa woodsii, Prosopis glandulosa, Phoradendron juniperinum, Syzygium claviflorum, Hyptis capitata, Mirabilis jalapa, and Ternstroemia gymnanthera are a few examples of foods and plants that contain it. It is composed of an aglycone component for triterpenoid saponins and a free acid group chemically and has historically been used for its cardiotonic, analgesic, anti-inflammatory, and hepatoprotective effects [53]. Oleanolic acid is known to have immune-modulatory effects by causing eukaryotic cells to release high mobility group box-1 protein (HMGB1) and macrophages to release C-reactive protein, endotoxins, and TNF-α. Through the secretion of pro-inflammatory cytokines such as IL-6, IL-1β, and TNF-α produced by macrophages, this extracellular HMGB1 acts as a potent immune stimulator [54]. Numerous immune-inflammatory diseases, including disseminated intravascular coagulation, atherosclerosis, sepsis, rheumatoid arthritis, xenotransplantation, and periodontitis, are known to have these immunological reactions. Oleanolic acid also prevents the LPS-induced activation of macrophage-like RAW264.7 cells by reducing the levels of HMGB1 proteins [55]. It is discovered to interact with a cyclooxygenase-2(COX2)-dependent mechanism to stimulate prostaglandin I2 (PGI2) release by human coronary smooth muscle cells [56]. Since it manages immunological problems, it is regarded as an immunomodulator.

Carotenoids are pigmented tetraterpenes, since they have strong light absorption and brilliant colour. They typically have a 40-carbon polyene chain and are made up of eight isoprene units. This nonradiative energy transfer mechanism enables them to absorb extra energy from other molecules [57]. Carotenoids are naturally occurring, lipid-soluble pigments that give host plants and animals their vibrant colour. Plant carotenoids may be crucial to maintaining human health [58]. They can act as potent antioxidants and are thought to treat several chronic illnesses, including cancer, osteoporosis, and cardiovascular disease. Some carotenoids, including lutein, β-carotene, and lycopene, may be able to reduce some inflammatory responses by modulating redox-sensitive signalling pathways like NF-kB and ROS [59–61]. The most prevalent cyclic tetraterpene and most potent pro-vitamin A in nature are β-carotene. It can be turned into vitamin A and is stored in the liver [57]. Lutein, a dihydroxy derivative of β-carotene and a standard component of many fruits and vegetables as well as egg yolks, is one of the lipophilic xanthophylls. It can prevent age-related macular degeneration, guard against oxidative stress, and have a neuroprotective impact on retinal inflammation [62, 63]. Lycopene, an additional acyclic tetraterpene, is the most prevalent carotenoid in the human body [58]. It is primarily found in fruits and vegetables that are red in colour. Since lycopene is a more potent antioxidant than vitamin E, it helps shield cells from free radical damage when there is oxidative stress. It has also been asserted that it lowers the chance of several chronic illnesses, including cardiovascular problems, RA, and atherosclerosis [57, 58]. These carotenoids with antioxidant properties may be developed into immunomodulators in the future.

2.3 Alkaloids

The most valuable and essential plant compounds are alkaloids, which are also powerful medicinal agents [64]. The alkaloid family is the largest class of secondary plant chemicals with one or more nitrogen atoms, typically combined as part of a cyclic structure, which consists of around 5500 identified compounds. At least one nitrogen atom with a basic nucleus is present [65]. Other elements found in alkaloids include carbon, hydrogen, oxygen, sulphur, chlorine, bromine, and phosphorus. These organic substances are neutral or only slightly acidic. A few synthetic substances have structure similar to natural alkaloids. Numerous creatures, including fungus, bacteria, mammals, and plants, contain alkaloids [66]. It has numerous medicinal effects, including anti-malarial, anti-asthmatic, anti-cancer, cholinomimetic, vasodilatory, anti-arrhythmic, analgesic, anti-bacterial, and anti-hyperglycemic effects. Several alkaloids have stimulant and psychoactive effects on the central nervous system (CNS) [67]. Additionally, it has immunomodulatory effects. Immune effector cells trigger the main immunomodulatory pathways of alkaloids to cause autoimmune responses [68]. It interacts with the forskolin proteins, lysosomes, phagocyte vacuoles, and neutrophil, monocyte, and macrophage cytoskeleton filaments. In addition, it triggers the innate immune response by causing macrophages to perform phagocytic functions [69]. Moreover, active antigen-presenting cells (APCs) i.e. macrophages, contribute to the production of adaptive immune responses [70]. Potent alkaloids i.e., berberine (BBR) and colchicine are identified as potential immunomodulatory drugs.

Berberine (BBR), an isoquinoline alkaloid, is present in several Berberis species. Its chemical name is 5,6-dihydro-9,10-dimethoxybenzo[g]-1,3-benzodioxolo[5,6-a] quinolizinium compound. It has anti-diabetic, hepatoprotective, hypolipidemic, cancer-preventive, anti-hypertensive, anti-oxidant, anti-inflammatory, anti-depressant, anti-diarrheal, and anti-microbial effects [71–73]. Multiple mechanisms of action are used to create these pharmacological effects. It interacts with multiple cellular kinases and signalling pathways. Some routes of the actions are interlinked with immunomodulatory pathways i.e., activation of nuclear factor erythroid-2-related factor-2 (Nrf2), MAPKs, NF-κB, & AMPK pathways; and expression of sirtuin 1 (SIRT1), & deacetylation of transcription factors of forkhead box O (FOXO) proteins [74]. In studies, BBR also lowers levels of proinflammatory cytokines such IFN-γ, IL-17, IL-6, and TNF-α, which results in immunomodulatory effects in non-obese diabetic (NOD) mice [73]. The overexpression of NADPH oxidase 2/4 is also downregulated, which results in antioxidant effects [75]. Furthermore, BBR stimulates Nrf2 activities, which increases the production of antioxidant enzymes such as NADPH quinine oxidoreductase-1 (NQO-1) and heme oxygenase-1 (HO-1). Phosphatidylinositol 3-kinase (PI3K)/protein kinase B (Akt), P38, and AMPK pathways are a few of the different cell signalling pathways that are promoted by Nrf2 [76]. As a result, it can be employed as an immunomodulatory agent to treat autoimmune diseases.

A significant natural alkaloid called colchicine is derived from the Colchicum autumnale plant (Colchicaceae family). Colchicine is a medication that has anti-inflammatory effects and is used to treat various inflammatory diseases including pericarditis, gout, and familial Mediterranean fever. It is known to impede microtubule polymerisation and reduce the growth of several types of cancer cells as its primary mechanism of action. Additionally, it has been discovered to decrease immunological responses such as neutrophil chemotaxis, lysosome breakdown, and leukocyte adhesiveness [77]. The immune cell signals i.e., nucleotide-binding and oligomerization domain (NOD), leucine-rich repeat proteins (LRR)-containing protein 3 and pyrin domain proteins [NLRP3 sensors] is suppressed by colchicine as well. SARS-CoV-2 proteins are also capable of initiating and energising comparable pathways. The therapy of COVID-19 with colchicine has tragically failed [78]. Moreover, it inhibits the production of leukotriene B4, neutrophil chemotactic factor and IL-1, which cause the immune cell regulatory action [79–81]. According to a literature review, colchicine can restrict procollagen synthesis, increase collagenase activity, and block mast cell release of histamine. It can also reduce the production of TNF-α that is caused by LPS [82–84]. Hence, autoimmune illnesses can be managed using their immunomodulatory effects.

2.4 Glycosides

The secondary metabolites of plants that contain sugar are called glycosides, a portion to which non-sugar portions are joined. The binding between the sugar and nonsugar moiety leads to hemiacetal formation. It includes the alcoholic or phenolic hydroxyl group of nonsugar and the aldehyde or keto group of the sugar moiety. These agents play numerous beneficial activities in animals and humans; however, many plants accumulate these chemicals in an inactive form which can be activated by the action of enzymes in the body [85]. Glycosides may be classified depending upon the glycone and aglycone moieties, such as glucoside, fructoside, α-glycosides and β-glycosides. Amygdalin and scrocaffeside-A are identified as primary bioactive immunomodulating agents. These substances primarily work to stimulate the immunological, cardiac, and central neurological systems. Furthermore, glycosides also show substantial antibacterial effects [86].

The bitter almond, apricot, plum, apple, and peach fruit kernels contain amygdalin, a cyanogenetic glycoside. Leukoderma, colorectal cancer, emphysema, leprosy, bronchitis, and asthma are all conditions frequently treated by amygdalin [87, 88]. By controlling T-cells, amygdalin has been shown to suppress inflammatory reactions and enhance immunomodulatory effects [89]. Additionally, it activates caspase-3, which suppresses Bcl-2-like protein 4 (Bax, proapoptotic protein) and B-cell lymphoma 2 (Bcl-2, an antiapoptotic protein) [88]. It stops the metastases of cancer cells via prevention of β1 and β4 integrins expression leading to suppression of the Akt-mediated mammalian target of rapamycin (mTOR) pathway (immunomodulatory mediators). Moreover, it lowers the β-catenin, integrin-linked kinase (ILK), and focal adhesion kinase (FAK) expression in immune cells [90]. Consequently, amygdalin is also an immunomodulatory agent.

Picrorhiza scrophulariiflora roots contain the caffeoyl glycoside known as scrocaffeside-A. It effectively heals leukoderma, inflammatory illnesses, gastrointestinal & urinary disorders, scorpion stings, and snake bites [91, 92]. Additionally, it controls the immunological responses of splenocytes by triggering concanavalin-A and LPS interactions [93]. According to an in vitro investigation, scrocaffeside-A increases CD4/CD8 population and cytokine production in splenocytes, which activates peritoneal macrophages and natural killer cell activity. Furthermore, it has been observed that scrocaffeside-A exposure increases the production of IFN-α, IL-2, IL-4, and IL-12 in cultured splenocytes. It suggests that scrocaffeside-A stimulates the host's immune system [94] and as a result, scrocaffeside-A is regarded as an immunomodulating substance.

2.5 Polysaccharides

Polycarbohydrates are another name for polysaccharides and are present in many foods. Its constituent monosaccharide units and glycosidic connections make up the long-chain polymeric carbohydrates. With the aid of amylase enzymes, it rapidly reacts with water through hydrolysis. This enzyme makes the constituent sugars i.e., monosaccharides, and oligosaccharides from polysaccharides [95]. Biologically, it is stored as starch, glycogen, galactogen, cellulose, and chitin. It is found in various plants including marine sources. Marine polysaccharides are used as medicine for variable disorders. Marine sources of polysaccharide have enriched contents of organic compounds like terpenoids, polyethers/ketides, lipo-glycoproteins, peptides, and polysaccharides [96]. It acts on various cell surface receptors altering cell proliferation and differentiation. Furthermore, it possesses immunomodulatory effects [97]. Pectin and Acemannan are also recognised as important bioactive immunomodulating substances.

The complex polysaccharide molecule known as pectin contains d-galacturonic acid monomers that have been esterified and connected by α-(1–4) chain [98]. In its typical state, pectin functions as an adsorbent and rapidly binds to various poisons, germs, and irritants in the intestinal mucosa. It lowers the pH in the intestinal lumen and has calming effects. The digestive system's alkaline environment is treated using modified pectin. Citrus pectin and modified citrus pectin, which are plant-based pectins, are known to have immunomodulatory effects by increasing the pro-inflammatory cytokines such as IFN-γ, IL-17, and TNF-α [99]. Pectin also inhibits the pro-inflammatory TLR2-TLR1 pathway, which is how it blocks the toll-like receptor 2 (TLR2) [100]. Specific polysaccharides produce immunomodulatory effects and influence immune cell function. Additionally, interactions between T-cells, monocytes, macrophages, and polymorphonuclear lymphocytes have altered innate and cell-mediated immunity through the influence of polymers of polysaccharides [101]. Because of this, it is utilised as an immunomodulatory medication to treat immunological diseases.

Acemannan, a mucopolysaccharide molecule, is widely distributed in aloe vera leaves. It has the chemical name β-(1,4)-acetylated soluble polymannose. It induces IFN, IL-1, TNF, and prostaglandin E2 release from activated macrophages. Additionally, it improves the control of macrophage phagocytosis, T-lymphocyte activity, and non-specific cytotoxicity. Potential anti-oxidant, antiviral, immunostimulant, antineoplastic, wound-healing, bone-proliferation, and neuroprotective effects are present [102]. It stimulates the generation of nitric oxide and macrophage-mannose receptors, which activate immune cells like macrophages [103]. Furthermore, through IFNγ-associated suppression of bcl-2 (B-cell lymphoma 2) expression, acemannan stimulates the RAW 264.7 cells [104]. Treat immunological disorders, it is regarded as an immunomodulatory agent.

2.6 Tannins

Tannins are substances with a high molecular weight that are water soluble and frequently found in plants as a complex with proteins, polysaccharides, and alkaloids. Depending upon their solubility or hydrolysis product, tannins are divided into hydrolysable tannins, proanthocyanidins, phlorotannins. Gallic acid esters are used to produce hydrolysable tannins. Phlorotannins are formed from phloroglucinol, obtained from brown algae, and condensed tannins are a combination of polyhydroxy flavan-3-ol monomers. Walnuts, peaches, berries, apples, and grapes are among the significant sources of tannins [105]. Numerous preclinical studies have shown their immunomodulatory properties.

Punicalagin (PCG), an ellagitannin, has several health benefits. According to Lee et al. investigation, the immunosuppressive properties of PCG derived from Punica granatum depend on its impact on the nuclear factor of triggered T cells (NFAT). Data showed administration of PCG inhibited leukocyte response, IL-2 expression, and CD3 + T cell infiltration. Moreover there is some evidence that PCG may be a free radical scavenger and thus can be used as potent immunosuppressive drug [106]. Reddy Reddana did yet another study on chebulagic acid's (CA) immunosuppressive properties, derived from Terminalia chebula, on LPS-induced RAW 264.7 cell line. The expression of IL-2, TNFα and ROS production was considerably reduced after treatment with CA. A dose-dependent trend was also observed in the inhibition of NF-κβ activation, p38, JNK, and ERK 1/2 phosphorylation [107]. Furthermore, Corilagin extracted from T. chebula showed the neuroprotective activity by downregulating the H_2O_2 stimulated PC12 cells death [108].

2.7 Saponins

The group of naturally occurring glycosides known as saponins is abundantly found in many different parts of plants, including leaves, flowers, shoots, roots, tubers, and seeds [109]. These are complicated compounds with a non-sugar (aglycone) component joined to a sugar moiety. Saponins fall into one of two groups based on their aglycone skeleton. Triterpens saponins, most of which are found in dicotyledonous angiosperms, make up the first class. Steroid saponins, which are primarily found in monotyledonous angiosperms, are found in the second class. Most of the oligosaccharides that make up the glycone portion of saponins are connected to the hydroxyl group by an acetal linkage. Numerous in vivo and in vitro investigations, have shown that plant-derived saponins can increase the immunogenicity of several vaccines. One of the most well-known functions of saponins is their usage as immunoadjuvants, which modulate the immune system produced by cells and aid in creating antibodies [109, 110]. Different saponin chemicals can stop the cell cycle, induce apoptosis and inhibit cancer cells. On rat liver microsomes, Ablise et al. examined the immunotherapeutic effect of glycyrrhizin produced from Glycyrrhiza glabra. With 1.0 mg/mL of glycyrrhizin, the classical complement pathway was significantly inhibited, and the antioxidant activity was increased. Another study by Punturee et al. found that utilising peripheral blood mononuclear cells (PBMCs) to extract Asiaticoside saponin from Centella asiatica had positive results and the data showed that as compared to the non-treated group, asiaticoside administration at 100 mg/kg significantly increased phagocytic index and total WBC count [111]. The immunological responses, both cellular and humoral, are also improved.

2.8 Sterols and sterolins

Combining sterols and sterolins improves NK cells' capacity to kill the NK 562 target cell line. Additionally, it has been proposed that specific ratios of sterols could restore the delicate balance between Th1 and Th2 cells, which decides how the immune response would turn out. At low concentrations, the phytosterols -Sitosterol and its glycoside more than doubled the in vitro proliferative response of T-cells triggered by sub-optimal amounts of phytohaemagglutinin. Moreover, it has been suggested that sterols and sterolins can control the amounts of Th1 and Th2 mediated cytokines, aiding in enhancing immune responses. Potent immunomodulators, phytosterols, β-Sitosterol, and its glycoside can enhance the proliferative responses of T cells even at low concentrations [112].

Rasool et al. used albino Wistar strain rats to study the immunomodulatory activities of withanolide derived from Withania somnifera. Withanolide administration in the rats dramatically reduced the proliferation of lymphocytes triggered by mitogens, the traditional complement pathway, and hypersensitive reactions. Withanolide might thus be developed into a potent immunosuppressive drug, according to the study [113]. Furthermore, by enhancing the Th1 and Th2 immune responses in mice with disseminated candidiasis, β-sitosterol and daucosterol also demonstrated immunomodulatory action [114]. The immunomodulatory properties of phytosterols isolated from Clinacanthus nutans by employing murine cells were described in another investigation by Lee and colleagues. To evaluate the immunosuppressive effects of phytosterols (stigmasterol, shaftoside, and β-sitosterol), mitogen-induced B and T-cell proliferation and the production of helper T-cell cytokines were observed. The results showed that treatment with phytosterols dramatically reduced T-cell proliferation and enhanced the production of Th1 and Th2 mediated cytokines [115].

The summary of phytocompounds and their mechanism of immunomodulation are expressed in **Tables 1–3**.

Compounds	Structures	Source	Mechanism of immunomodulation	References
Myricetin	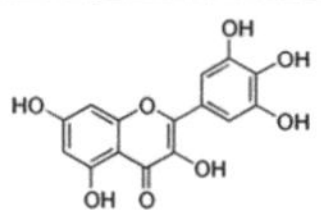	*Fragaria ananassa Spinacia oleracea*	Inhibits IL-12 production by reducing NF-κB binding activity.	[116]
Luteolin	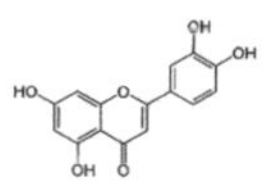	*Lonicera japonica*	Suppresses LPS-mediated protein kinase B and IKK phosphorylation and ROS production.	[117]
Chrysin	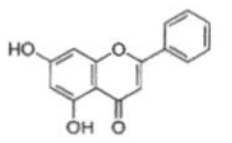	Honey, propolis, blue passion flower	Inhibits prostaglandin-E2, COX-2 and NF-κB.	[118]
Baicalein	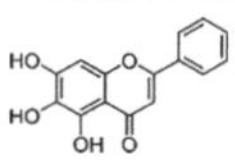	*Scutellaria baicalensis georgi*	Suppress inflammatory cytokines production.	[119]
Vitexin	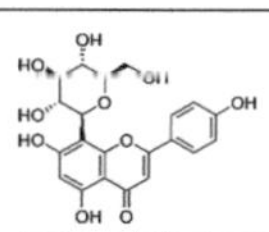	*Vitex lucens Kirk*	Inhibits leukocyte migration.	[120]
Seselin	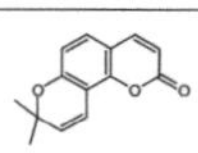	*Plumbago zeylanica*	Reduces proinflammatory factors level and activity of STAT-1 and p65.	[121]
Angelicin	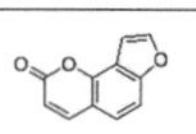	*Psoralea corylifolia*	Inhibits phosphorylation of NF-κB, p65, p38, MAPK and JNK in LPS-induced acute lung injury.	[122]

Table 1.
Immunomodulatory actions of flavanoids and coumarins.

Compounds	Structures	Source	Mechanism of immunomodulation	References
Piperine	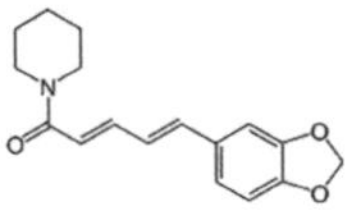	*Rhododendrone faurie*	Inhibition of IgM antibody secretion	[123]
		Vicoa indica Anethum sowa	Upregulation of IL-10 and NF-κB	
Sophocarpine	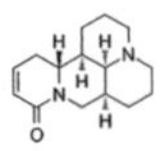	*Sophora alopecuroides L.*	Reduces NO, TNF-α, IL-6 production	[124]
			Suppress the expression of iNOS and COX-2	
Koumine	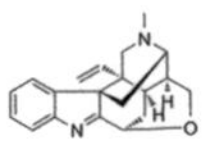	*Gelsemium elegans*	Increased the proportion of $CD4^+/CD8^+$ T cells Reduces the percentage of Th1 and Th17 cells Increases Th2 and Treg cells	[125]
Lycorine	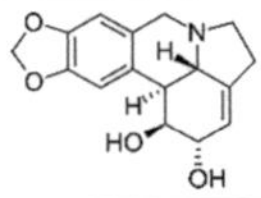	*Lycoris radiate*	Decreases phosphorylation of Bc1–2 by the activation of MEK–ERK.	[126]
Ursolic acid	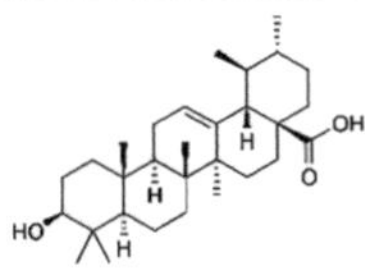	*Pimpinella major Gladiolus italicus*	Suppresses lymphocyte proliferation. Inhibits cytokine secretion.	[127]
Tryptolide	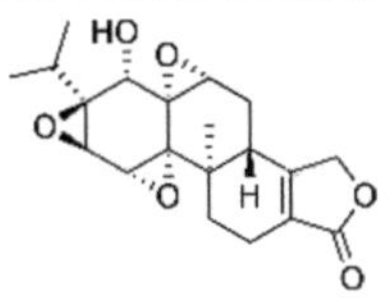	*Trypterigium wilfordii*	Inhibits IL-6/STAT-3 pathways. Suppress toll like receptor protein2/4 and altered MAPK, p38 and ERK pathways.	[128]
Echinocystic acid	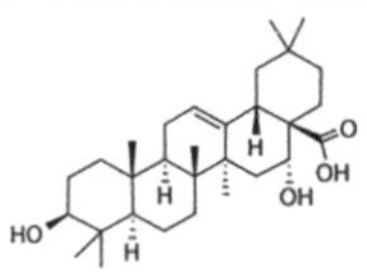	*Gleditsia sinensis*	Induces NF-κB activation and ERK phosphorylation.	[129]
Celastrol	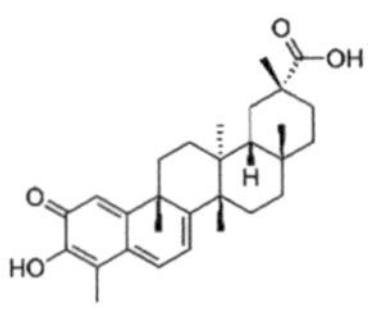	*Trypterigium wilfordii*	Suppresses TNF-α induced production of anti-apoptosis proteins.	[130]

Table 2.
Immunomodulatory actions of alkaloids and terpenoids.

Compounds	Structures	Source	Mechanism of immunomodulation	References
Chebugalic acid	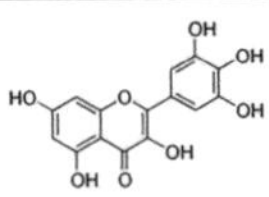	*Terminalia chebula*	Increase lymphocyte proliferation. Suppresses ROS production.	[131]
Punicalagin	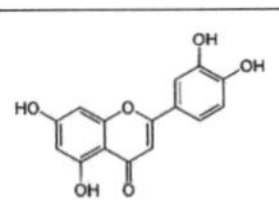	*Punica granatum*	Decrease IL-2 production. Suppresses leukocyte reaction.	[106]

Compounds	Structures	Source	Mechanism of immunomodulation	References
Corilagin	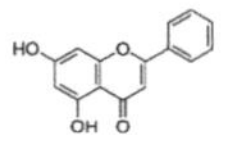	*Phyllanthus winaria*	Inhibits NF-κB activity.	[132]
Mangiferin	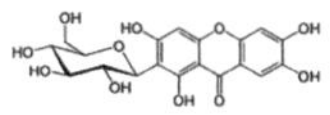	*Mangifera indica*	Inhibits IFN-γ production.	[133]
Aucubin	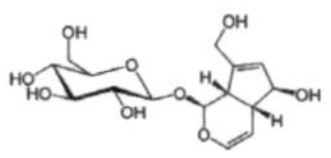	*Aucuba japonica*	Inhibits NO production. Decreases release of inflammatory mediators. Suppresses neutrophils and macrophages	[134]
Asiaticoside	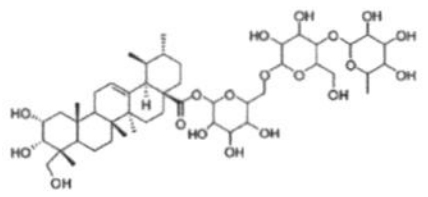	*Centella asiatica*	Increased IL-2 and TNF-α production. Induces NO production.	[135]
Glycyrrhizin	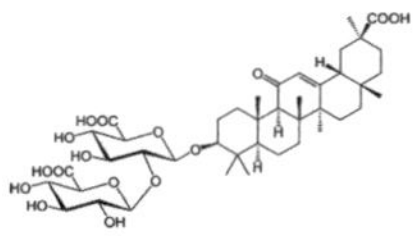	Liquorice	Increases IL-2 production.	[136]

Table 3.
Immunomodulatory actions of tannins, glycosides and saponins.

3. Conclusion

Due to their diverse pharmacological properties, active components in medicinal plants have been demonstrated as an essential source of clinical medicines. Traditional medicine uses medicinal plants for health benefits and is thoroughly researched. Any disparity between immune system causes numerous fatal diseases, including cancer, rheumatoid arthritis, and diabetes. According to several recent studies, these disorders can be treated by phytochemicals, present in particular plant species. In vivo, in vitro and ex vivo studies have reported that the administration of phytochemicals such as quercetin, ellagic acid, mangiferin and withanolide, significantly reduced the occurrence of immune-related diseases due to their antioxidant, anti-inflammatory and immunomodulatory activities. Additionally, phytochemicals frequently enhance the Th1 as well as Th2-mediated cytokine production that decreases autoimmune conditions. Despite numerous preclinical and ex vivo studies, other practical proof is needed to support the immunomodulatory action of phytochemicals in preventing and treating immune system-related illnesses. Specific parameter enhancements, like standardised dosage and duration of intervention, are still needed to create the ideal phytochemical immunomodulators.

Acknowledgements

Authors extend a deep sense of gratitude to Hon'ble Vice Chancellor, SRM University, Kattankulathur, Chengalpettu district, Tamil Nadu, India.

Conflict of interest

The authors declare no conflict of interest.

Author details

Ayda Cherian and Velmurugan Vadivel*
SRM College of Pharmacy, SRMIST, Kattankulathur, India

*Address all correspondence to: velmuruv@srmist.edu.in

References

[1] Webb NE, Bernshtein B, Alter G. Tissues: The unexplored frontier of antibody mediated immunity. Current Opinion in Virology. 2021;**47**:52-67. DOI: 10.1016/j.coviro.2021.01.001

[2] Janeway Jr CA, Travers P, Walport M, Shlomchik MJ. Principles of innate and adaptive immunity. In: Immunobiology: The Immune System in Health and Disease. 5th ed. New York: Garland Science; 2001

[3] Tomasello G, Tralongo P, Damiani P, Sinagra E, Di Trapani B, Zeenny MN, et al. Dismicrobism in inflammatory bowel disease and colorectal cancer: Changes in response of colocytes. World journal of gastroenterology: WJG. 2014; **20**(48):18121. DOI: 10.3748/wjg.v20.i48.18121

[4] Chastin SF, Abaraogu U, Bourgois JG, Dall PM, Darnborough J, Duncan E, et al. Effects of regular physical activity on the immune system, vaccination and risk of community-acquired infectious disease in the general population: Systematic review and meta-analysis. Sports Medicine. 2021;**51**:1673-1686. DOI: 10.1007/s40279-021-01466-1

[5] Abou-Raya A, Abou-Raya S. Inflammation: A pivotal link between autoimmune diseases and atherosclerosis. Autoimmunity Reviews. 2006;**5**(5):331-337. DOI: 10.1016/j.autrev.2005.12.006

[6] Nair A, Chattopadhyay D, Saha B. Plant-derived immunomodulators. In: New look to phytomedicine. Cambridge, Massachusetts: Academic Press; 2019. pp. 435-499. DOI: 10.1016/b978-0-12-814619-4.00018-5

[7] Parkin J, Cohen B. An overview of the immune system. The Lancet. 2001; **357**(9270):1777-1789. DOI: 10.1016/S0140-6736(00)04904-7

[8] Adusei KM, Ngo TB, Sadtler K. T lymphocytes as critical mediators in tissue regeneration, fibrosis, and the foreign body response. Acta Biomaterialia. 2021;**133**:17-33. DOI: 10.1016/j.actbio.2021.04.023

[9] Turvey SE, Broide DH. Innate immunity. Journal of Allergy and Clinical Immunology. 2010;**125**(2):S24-S32. DOI: 10.1016/j.jaci.2009.07.016

[10] Sarma JV, Ward PA. The complement system. Cell and tissue research. 2011;**343**(1):227-235. DOI: 10.1007/s00441-010-1034-0

[11] Takeuchi O, Akira S. Pattern recognition receptors and inflammation. Cell. 2010;**140**(6):805-820. DOI: 10.1016/j.cell.2010.01.022

[12] Toebak MJ, Gibbs S, Bruynzeel DP, Scheper RJ, Rustemeyer T. Dendritic cells: Biology of the skin. Contact Dermatitis. 2009;**60**(1):2-0. DOI: 10.1111/j.1600-0536.2008.01443.x

[13] Spellberg B, Edwards JE Jr. Type 1/type 2 immunity in infectious diseases. Clinical Infectious Diseases. 2001;**32**(1): 76-102. DOI: 10.1086/317537

[14] Hengge UR, Benninghoff B, Ruzicka T, Goos M. Topical immunomodulators—Progress towards treating inflammation, infection, and cancer. The Lancet infectious diseases. 2001;**1**(3):189-198. DOI: 10.1016/S1473-3099(01)00095-0

[15] Shukla S, Bajpai VK, Kim M. Plants as potential sources of natural immunomodulators. Reviews in

Environmental Science and Bio/ Technology. 2014;**13**(1):17-33. DOI: 10.1007/s11157-012-9303-x

[16] Alhazmi HA, Najmi A, Javed SA, Sultana S, Al Bratty M, Makeen HA, et al. Medicinal plants and isolated molecules demonstrating immunomodulation activity as potential alternative therapies for viral diseases including COVID-19. Frontiers in Immunology. 2021;**12**:1721. DOI: 10.3389/fimmu.2021.637553

[17] Kharkar PB, Talkar SS, Kadwadkar NA, Patravale VB. Nanosystems for oral delivery of immunomodulators. Nanostructures for Oral Medicine. Amsterdam, Netherlands: Elsevier; 2017. pp. 295-334. DOI: 10.1016/B978-0-323-47720-8.00012-2

[18] Israeli E, Agmon-Levin N, Blank M, Shoenfeld Y. Adjuvants and autoimmunity. Lupus. 2009;**18**(13): 1217-1225. DOI: 10.1177/09612033 09345724

[19] Yasrib Q, Abid H, Shashank KS, Ajit KS. Potential role of natural molecules in health and disease: Importance of boswellic acid. Journal of Medicinal Plants Research. 2010;**4**(25): 2778-2785

[20] Licciardi PV, Underwood JR. Plant-derived medicines: A novel class of immunological adjuvants. International Immunopharmacology. 2011;**11**(3): 390-398. DOI: 10.1016/j.intimp.2010. 10.014

[21] Cione E, La Torre C, Cannataro R, Caroleo MC, Plastina P, Gallelli L. Quercetin, epigallocatechin gallate, curcumin, and resveratrol: From dietary sources to human microRNA modulation. Molecules. 2019;**25**(1):63. DOI: 10.3390/molecules25010063

[22] Zhao Y, Roy S, Wang C, Goel A. A combined treatment with Berberine and Andrographis exhibits enhanced anti-cancer activity through suppression of DNA replication in colorectal cancer. Pharmaceuticals. 2022;**15**(3):262. DOI: 10.3390/ph15030262

[23] Bain BJ. Structure and function of red and white blood cells and platelets. Medicine. 2021;**49**:183-188. DOI: 10.1016/j.mpmed.2021.01.001

[24] Zeinali M, Rezaee SA, Hosseinzadeh H. An overview on immunoregulatory and anti-inflammatory properties of chrysin and flavonoids substances. Biomedicine & Pharmacotherapy. 2017;**92**:998-1009. DOI: 10.1016/j.biopha.2017.06.003

[25] Ramadass V, Vaiyapuri T, Tergaonkar V. Small molecule NF-κB pathway inhibitors in clinic. International journal of molecular sciences. 2020;**21**(14):5164. DOI: 10.3390/ijms21145164

[26] Sivagami B, Sailaja B. A review on analytical methods for antiviral Phytoconstituents. Journal of Young Pharmacists. 2021;**13**(1):7-13. DOI: 10.5530/jyp.2021.13.2

[27] Abbas M, Saeed F, Anjum FM, Afzaal M, Tufail T, Bashir MS, et al. Natural polyphenols: An overview. International Journal of Food Properties. 2017;**20**(8):1689-1699. DOI: 10.1080/ 10942912.2016.1220393

[28] Lamoral-Theys D, Pottier L, Dufrasne F, Neve J, Dubois J, Kornienko A, et al. Natural polyphenols that display anticancer properties through inhibition of kinase activity. Current medicinal chemistry. 2010; **17**(9):812-825. DOI: 10.2174/0929867 10790712183

[29] Bhat KP, Kosmeder JW, Pezzuto JM. Biological effects of resveratrol. Antioxidants and redox signaling. 2001; **3**(6):1041-1064. DOI: 10.1089/152308601317203567

[30] Subbaramaiah K, Chung WJ, Michaluart P, Telang N, Tanabe T, Inoue H, et al. Resveratrol inhibits cyclooxygenase-2 transcription and activity in phorbol ester-treated human mammary epithelial cells. Journal of Biological Chemistry. 1998;**273**(34): 21875-21882. DOI: 10.1074/jbc.273.34.21875

[31] Manna SK, Mukhopadhyay A, Aggarwal BB. Resveratrol suppresses TNF-induced activation of nuclear transcription factors NF-κB, activator protein-1, and apoptosis: Potential role of reactive oxygen intermediates and lipid peroxidation. The Journal of Immunology. 2000;**164**(12):6509-6519. DOI: 10.4049/jimmunol.164.12.6509

[32] Holmes-McNary M, Baldwin AS Jr. Chemopreventive properties of trans-resveratrol are associated with inhibition of activation of the IκB kinase. Cancer Research. 2000;**60**(13):3477-3483

[33] Das S, Das DK. Anti-inflammatory responses of resveratrol. Inflammation & Allergy-Drug Targets (Formerly Current Drug Targets-Inflammation & Allergy). 2007;**6**(3):168-173. DOI: 10.2174/187152807781696428

[34] Choo QY, Yeo SC, Ho PC, Tanaka Y, Lin HS. Pterostilbene surpassed resveratrol for anti-inflammatory application: Potency consideration and pharmacokinetics perspective. Journal of Functional Foods. 2014;**11**:352-362. DOI: 10.1016/j.jff.2014.10.018

[35] Cichocki M, Paluszczak J, Szaefer H, Piechowiak A, Rimando AM, Baer-Dubowska W. Pterostilbene is equally potent as resveratrol in inhibiting 12-O-tetradecanoylphorbol-13-acetate activated NFκB, AP-1, COX-2, and iNOS in mouse epidermis. Molecular Nutrition & Food Research. 2008;**52**(S1):S62-S70. DOI: 10.1002/mnfr.200700466

[36] Alam MA, Subhan N, Hossain H, Hossain M, Reza HM, Rahman MM, et al. Hydroxycinnamic acid derivatives: A potential class of natural compounds for the management of lipid metabolism and obesity. Nutrition and Metabolism. 2016;**13**:1-13. DOI: 10.1186/s12986-016-0080-3

[37] Taofiq O, González-Paramás AM, Barreiro MF, Ferreira IC. Hydroxycinnamic acids and their derivatives: Cosmeceutical significance, challenges and future perspectives, a review. Molecules. 2017;**22**(2):281. DOI: 10.3390/molecules22020281

[38] Vaughn AR, Haas KN, Burney W, Andersen E, Clark AK, Crawford R, et al. Potential role of curcumin against biofilm-producing organisms on the skin: A review. Phytotherapy Research. 2017;**31**(12):1807-1816. DOI: 10.1002/ptr.5912

[39] Mehdi S, Siddique R, Mehdi S, Ali MM. Phytochemical evaluation of curcuma longa and its beneficial effects. Journal of Toxicological & Pharmaceutical Sciences. 2021;**5**(1):5-10

[40] Shishodia S. Molecular mechanisms of curcumin action: Gene expression. BioFactors. 2013;**39**:37-55. DOI: 10.1002/biof.1041

[41] Liu S, Wang Z, Hu Z, Zeng X, Li Y, Su Y, et al. Anti-tumor activity of curcumin against androgen-independent prostate cancer cells via inhibition of NF-κB and AP-1 pathway in vitro. Journal of Huazhong University of

Science and Technology [Medical Sciences]. 2011;**31**(4):530-534. DOI: 10.1007/s11596-011-0485-1

[42] Ververidis F, Trantas E, Douglas C, Vollmer G, Kretzschmar G, Panopoulos N. Biotechnology of flavonoids and other phenylpropanoid-derived natural products. Part I: Chemical diversity, impacts on plant biology and human health. Biotechnology journal: Healthcare nutrition. Technology. 2007;**2**(10): 1214-1234. DOI: 10.1002/biot.20070 0084

[43] Gomes A, Fernandes E, Lima JL, Mira L, Corvo ML. Molecular mechanisms of anti-inflammatory activity mediated by flavonoids. Current medicinal chemistry. 2008;**15**(16): 1586-1605. DOI: 10.2174/09298670 8784911579

[44] David AV, Arulmoli R, Parasuraman S. Overviews of biological importance of quercetin: A bioactive flavonoid. Pharmacognosy Reviews. 2016;**10**(20):84. DOI: 10.4103/0973-7847.194044

[45] Brown MD. Green tea (Camellia sinensis) extract and its possible role in the prevention of cancer. Alternative medicine review: a journal of clinical therapeutic. 1999;**4**(5):360-370

[46] Conforti F, Menichini F. Phenolic compounds from plants as nitric oxide production inhibitors. Current medicinal chemistry. 2011;**18**(8): 1137-1145. DOI: 10.2174/ 092986711795029690

[47] Zhang C, Hong K. Production of terpenoids by synthetic biology approaches. Frontiers in bioengineering and biotechnology. 2020;**8**:347. DOI: 10.3389/fbioe.2020.00347

[48] Grassman J. Terpenoids as plant antioxidants. Vitamins & Hormones. 2005;**72**:505-535. DOI: 10.1016/ S0083-6729(05)72015-X

[49] Plata-Rueda A, Campos JM, da Silva RG, Martínez LC, Dos Santos MH, Fernandes FL, et al. Terpenoid constituents of cinnamon and clove essential oils cause toxic effects and behavior repellency response on granary weevil. Sitophilus granarius. Ecotoxicology and environmental safety. 2018;**156**:263-270. DOI: 10.1016/ j.ecoenv.2018.03.033

[50] Chiou WF, Chen CF, Lin JJ. Mechanisms of suppression of inducible nitric oxide synthase (iNOS) expression in RAW 264.7 cells by andrographolide. British Journal of Pharmacology. 2000; **129**(8):1553-1560. DOI: 10.1038/sj. bjp.0703191

[51] Maiti K, Gantait A, Kakali M, Saha BP, Mukherjee PK. Therapeutic potentials of andrographolide from Andrographis paniculata: A review. Journal of Natural Remedies. 2006;**6**: 1-13. DOI: 10.18311/jnr/2006/272

[52] Ding Y, Chen L, Wu W, Yang J, Yang Z, Liu S. Andrographolide inhibits influenza a virus-induced inflammation in a murine model through NF-κB and JAK-STAT signaling pathway. Microbes and infection. 2017;**19**(12):605-615. DOI: 10.1016/j.micinf.2017.08.009

[53] Liu J. Pharmacology of oleanolic acid and ursolic acid. Journal of Ethnopharmacology. 1995;**49**(2):57-68. DOI: 10.1016/0378-8741(95)90032-2

[54] Han Y, Tong Z, Wang C, Li X, Liang G. Oleanolic acid exerts neuroprotective effects in subarachnoid hemorrhage rats through SIRT1-mediated HMGB1 deacetylation. European Journal of Pharmacology.

2021;**893**:173811. DOI: 10.1016/j.ejphar.2020.173811

[55] Kawahara KI, Hashiguchi T, Masuda K, Saniabadi AR, Kikuchi K, Tancharoen S, et al. Mechanism of HMGB1 release inhibition from RAW264. 7 cells by oleanolic acid in Prunus mume Sieb. Et Zucc. International journal of molecular medicine. 2009;**23**(5):615-620. DOI: 10.3892/ijmm_00000172

[56] Martinez-Gonzalez J, Rodriguez-Rodriguez R, Gonzalez-Diez M, Rodriguez C, Herrera MD, Ruiz-Gutierrez V, et al. Oleanolic acid induces prostacyclin release in human vascular smooth muscle cells through a cyclooxygenase-2-dependent mechanism. The Journal of nutrition. 2008;**138**(3):443-448. DOI: 10.1093/jn/138.3.443

[57] Pan MH, Ho CT. Chemopreventive effects of natural dietary compounds on cancer development. Chemical Society Reviews. 2008;**37**(11):2558-2574. DOI: 10.1039/B801558A

[58] Salminen A, Lehtonen M, Suuronen T, Kaarniranta K, Huuskonen J. Terpenoids: Natural inhibitors of NF-κB signaling with anti-inflammatory and anticancer potential. Cellular and Molecular Life Sciences. 2008;**65**(19):2979-2999. DOI: 10.1007/s00018-008-8103-5

[59] Chew BP, Park JS. Carotenoid action on the immune response. The Journal of nutrition. 2004;**134**(1):257S-261S. DOI: 10.1093/jn/134.1.257S

[60] De Stefano D, Maiuri MC, Simeon V, Grassia G, Soscia A, Cinelli MP, et al. Lycopene, quercetin and tyrosol prevent macrophage activation induced by gliadin and IFN-γ. European journal of pharmacology. 2007;**566**(1–3):192-199. DOI: 10.1016/j.ejphar.2007.03.051

[61] Huang CS, Fan YE, Lin CY, Hu ML. Lycopene inhibits matrix metalloproteinase-9 expression and down-regulates the binding activity of nuclear factor-kappa B and stimulatory protein-1. The Journal of nutritional biochemistry. 2007;**18**(7):449-456. DOI: 10.1016/j.jnutbio.2006.08.007

[62] Lee EH, Faulhaber D, Hanson KM, Ding W, Peters S, Kodali S, et al. Dietary lutein reduces ultraviolet radiation-induced inflammation and immunosuppression. Journal of Investigative Dermatology. 2004;**122**(2):510-517. DOI: 10.1046/j.0022-202X.2004.22227.x

[63] Sasaki M, Ozawa Y, Kurihara T, Noda K, Imamura Y, Kobayashi S, et al. Neuroprotective effect of an antioxidant, lutein, during retinal inflammation. Investigative ophthalmology & visual science. 2009;**50**(3):1433-1439. DOI: 10.1167/iovs.08-2493

[64] Okwu DE. Phytochemicals, vitamins and mineral contents of two Nigerian medicinal plants. International Journal of Molecular Medicine and Advance Sciences. 2005;**1**(4):375-381

[65] Harborne JB. Phytochemical Methods. London: Chapman and Hall, Ltd.; 1973. pp. 49-188

[66] Badri S, Basu VR, Chandra K, Anasuya D. A review on pharmacological activities of alkaloids. World Journal of Current Medical and Pharmaceutical Research. 2019;**1**:230-234. DOI: 10.37022/WJCMPR.2019.01068

[67] Qin N, Lu X, Liu Y, Qiao Y, Qu W, Feng F, et al. Recent research progress of Uncaria spp. based on alkaloids:

Phytochemistry, pharmacology and structural chemistry. European Journal of Medicinal Chemistry. 2021;**210**: 112960. DOI: 10.1016/j.ejmech.2020.112960

[68] Jacques AS, Arnaud SSS, Fréjus OOH, Jacques DT. Review on biological and immunomodulatory properties of Moringa oleifera in animal and human nutrition. Journal of Pharmacognosy and Phytotherapy. 2020;**12**:1-9. DOI: 10.5897/jpp2019.0551

[69] Yakubu Y, Talba AM, Chong CM, Ismail IS, Shaari K. Effect of Terminalia catappa methanol leaf extract on nonspecific innate immune responses and disease resistance of red hybrid tilapia against Streptococcus agalactiae. Aquaculture Reports. 2020;**18**:100555. DOI: 10.1016/j.aqrep.2020.100555

[70] Ackermann M, Dragon AC, Lachmann N. The immune-modulatory properties of iPSC-derived antigen-presenting cells. Transfusion Medicine and Hemotherapy. 2020;**47**(6):444-453. DOI: 10.1159/000512721

[71] Neag MA, Mocan A, Echeverría J, Pop RM, Bocsan CI, Crişan G, et al. Berberine: Botanical occurrence, traditional uses, extraction methods, and relevance in cardiovascular, metabolic, hepatic, and renal disorders. Frontiers in pharmacology. 2018;**9**:557. DOI: 10.3389/fphar.2018.00557

[72] Imanshahidi M, Hosseinzadeh H. Pharmacological and therapeutic effects of Berberis vulgaris and its active constituent, berberine. Phytotherapy Research. 2008;**22**(8):999-1012. DOI: 10.1002/ptr.2399

[73] Li Z, Geng YN, Jiang JD, Kong WJ. Antioxidant and anti-inflammatory activities of berberine in the treatment of diabetes mellitus. Evidence-based complementary and alternative medicine. 2014;**2014**:1-12. DOI: 10.1155/2014/289264

[74] Ma X, Chen Z, Wang L, Wang G, Wang Z, Dong X, et al. The pathogenesis of diabetes mellitus by oxidative stress and inflammation: Its inhibition by berberine. Frontiers in Pharmacology. 2018;**9**:782. DOI: 10.3389/fphar.2018.00782

[75] Ghorbani N, Sahebari M, Mahmoudi M, Rastin M, Zamani S, Zamani M. Berberine inhibits the gene expression and production of proinflammatory cytokines by mononuclear cells in rheumatoid arthritis and healthy individuals. Current Rheumatology Reviews. 2021;**17**(1): 113-121. DOI: 10.2174/1573397116666200907111303

[76] Qin S, Tang H, Li W, Gong Y, Li S, Huang J, et al. AMPK and its activator berberine in the treatment of neurodegenerative diseases. Current Pharmaceutical Design. 2020;**26**(39): 5054-5066. DOI: 10.2174/1381612826666200523172334

[77] Schwarz YA, Kivity S, Ilfeld DN, Schlesinger M, Greif J, Topilsky M, et al. A clinical and immunologic study of colchicine in asthma. Journal of allergy and clinical immunology. 1990;**85**(3): 578-582. DOI: 10.1016/0091-6749(90)90096-M

[78] Gendelman O, Amital H, Bragazzi NL, Watad A, Chodick G. Continuous hydroxychloroquine or colchicine therapy does not prevent infection with SARS-CoV-2: Insights from a large healthcare database analysis. Autoimmunity reviews. 2020;**19**(7):102566. DOI: 10.1016/j.autrev.2020.102566

[79] Peters SP, Freeland HS, Kelly SJ, Pipkorn U, Naclerio RM, Proud D, et al.

Is leukotriene B4 an important mediator in human IgE-mediated allergic reactions? American Review of Respiratory Disease. 1987;**135**(6P2): S42-S45

[80] Spilberg I, Mandell B, Mehta J, Simchowitz L, Rosenberg D. Mechanism of action of colchicine in acute urate crystal-induced arthritis. The Journal of clinical investigation. 1979;**64**(3): 775-780. DOI: 10.1172/JCI109523

[81] Kershenobich D, Rojkind M, Quiroga A, Alcocer-Varela J. Effect of colchicine on lymphocyte and monocyte function and its relation to fibroblast proliferation in primary biliary cirrhosis. Hepatology. 1990;**11**(2):205-209. DOI: 10.1002/hep.1840110208

[82] Fell HB, Lawrence CE, Bagga MR, Hembry RM, Reynolds JJ. The degradation of collagen in pig synoviurn in vitro and the effect of colchicine. Matrix. 1989;**9**(2):116-126. DOI: 10.1016/ S0934-8832(89)80029-0

[83] Gillespie E, Levine RJ, Malawista SE. Histamine release from rat peritoneal mast cells: Inhibition by colchicine and potentiation by deuterium oxide. Journal of Pharmacology and Experimental Therapeutics. 1968;**164**(1):158-165

[84] Li Z, Davis GS, Mohr C, Nain M, Gemsa D. Suppression of LPS-induced tumor necrosis factor-α gene expression by microtubule disrupting agents. Immunobiology. 1996;**195**(4–5): 640-654. DOI: 10.1016/S0171-2985(96) 80028-3

[85] Brito-Arias M. Nucleoside mimetics. In: Synthesis and Characterization of Glycosides. Boston, MA: Springer; 2007. pp. 179-246. DOI: 10.1007/978-0-387-70792-1_4

[86] Nenaah G. Antimicrobial activity of Calotropis procera Ait.(Asclepiadaceae) and isolation of four flavonoid glycosides as the active constituents. World Journal of Microbiology and Biotechnology. 2013;**29**(7):1255-1262. DOI: 10.1007/ s11274-013-1288-2

[87] Gleadow RM, Møller BL. Cyanogenic glycosides: Synthesis, physiology, and phenotypic plasticity. Annual Review of Plant Biology. 2014;**65**(1):155-185. DOI: 10.1146/annurev-arplant-050213-040027

[88] Chang HK, Shin MS, Yang HY, Lee JW, Kim YS, Lee MH, et al. Amygdalin induces apoptosis through regulation of Bax and Bcl-2 expressions in human DU145 and LNCaP prostate cancer cells. Biological and Pharmaceutical Bulletin. 2006;**29**(8): 1597-1602. DOI: 10.1248/bpb.29.1597

[89] Jiagang D, Li C, Wang H, Hao E, Du Z, Bao C, et al. Amygdalin mediates relieved atherosclerosis in apolipoprotein E deficient mice through the induction of regulatory T cells. Biochemical and Biophysical Research Communications. 2011;**411**(3):523-529. DOI: 10.1016/j.bbrc.2011.06.162

[90] Qian L, Xie B, Wang Y, Qian J. Amygdalin-mediated inhibition of non-small cell lung cancer cell invasion in vitro. International journal of clinical and experimental pathology. 2015;**8**(5): 5363

[91] Zhu TF, Huang KY, Deng XM, Zhang Y, Xiang H, Gao HY, et al. Three new caffeoyl glycosides from the roots of Picrorhiza scrophulariiflora. Molecules. 2008;**13**:729-735. DOI: 10.3390/ molecules13040729

[92] Dey AC. Indian Medicinal Plants Used in Ayurvedic Preparations. Dehradun: Bishen Singh Mahendra Pal Singh; 1980

[93] Gurjar VK, Pal D. Natural compounds extracted from medicinal plants and their immunomodulatory activities. Bioactive Natural Products for Pharmaceutical Applications. 2021;**140**: 197-261. DOI: 10.1007/978-3-030-54027-2_6

[94] An N, Wang D, Zhu T, Zeng S, Cao Y, Cui J, et al. Effects of scrocaffeside a from Picrorhiza Scrophulariiflora on immunocyte function in vitro. Immunopharmacology and Immunotoxicology. 2009;**31**(3): 451-458. DOI: 10.1080/08923970902783092

[95] Hou C, Chen L, Yang L, Ji X. An insight into anti-inflammatory effects of natural polysaccharides. International journal of biological macromolecules. 2020;**153**:248-255. DOI: 10.1016/j.ijbiomac.2020.02.315

[96] Zhong R, Wan X, Wang D, Zhao C, Liu D, Gao L, et al. Polysaccharides from marine Enteromorpha: Structure and function. Trends in Food Science & Technology. 2020; **99**:11-20. DOI: 10.1016/j.tifs.2020.02.030

[97] Barbosa JR, de Carvalho Junior RN. Polysaccharides obtained from natural edible sources and their role in modulating the immune system: Biologically active potential that can be exploited against COVID-19. Trends in Food Science & Technology. 2021; **108**:223-235. DOI: 10.1016/j.tifs.2020.12.026

[98] Van Buggenhout S, Sila DN, Duvetter T, Van Loey A, Hendrickx MJ. Pectins in processed fruits and vegetables: Part III—Texture engineering. Comprehensive reviews in food science and food safety. 2009;**8**(2): 105-117. DOI: 10.1111/j.1541-4337.2009.00072.x

[99] Merheb R, Abdel-Massih RM, Karam MC. Immunomodulatory effect of natural and modified citrus pectin on cytokine levels in the spleen of BALB/c mice. International journal of biological macromolecules. 2019;**121**:1-5. DOI: 10.1016/j.ijbiomac.2018.09.189

[100] Sahasrabudhe NM, Beukema M, Tian L, Troost B, Scholte J, Bruininx E, et al. Dietary fiber pectin directly blocks toll-like receptor 2–1 and prevents doxorubicin-induced ileitis. Frontiers in immunology. 2018;**9**:383. DOI: 10.3389/fimmu.2018.00383

[101] Tzianabos AO. Polysaccharide immunomodulators as therapeutic agents: Structural aspects and biologic function. Clinical Microbiology Reviews. 2000;**13**:523-533. DOI: 10.1128/CMR.13.4.523

[102] Liu C, Cui Y, Pi F, Cheng Y, Guo Y. Qian H extraction, purification, structural characteristics, biological activities and pharmacological applications of acemannan, a polysaccharide from aloe vera: A review. Molecules. 2019;**24**:1554. DOI: 10.3390/molecules24081554

[103] Karaca K, Sharma JM, Nordgren R. Nitric oxide production by chicken macrophages activated by Acemannan, a complex carbohydrate extracted from Aloe vera. International Journal of Immunopharmacology. 1995;**17**(3): 183-188. DOI: 10.1016/0192-0561(94)00102-T

[104] Ramamoorthy L, Tizard IR. Induction of apoptosis in a macrophage cell line RAW 264.7 by acemannan, a β-(1, 4)-acetylated mannan. Molecular Pharmacology. 1998;**53**(3):415-421

[105] Serrano J, Puupponen-Pimiä R, Dauer A, Aura AM, Saura-Calixto F.

Tannins: Current knowledge of food sources, intake, bioavailability and biological effects. Molecular nutrition & food research. 2009;**53**(S2):S310-S329. DOI: 10.1002/mnfr.200900039

[106] Lee SI, Kim BS, Kim KS, Lee S, Shin KS, Lim JS. Immune-suppressive activity of punicalagin via inhibition of NFAT activation. Biochemical and biophysical research communications. 2008;**371**(4):799-803. DOI: 10.1016/j.bbrc.2008.04.150

[107] Reddy DB, Reddanna P. Chebulagic acid (CA) attenuates LPS-induced inflammation by suppressing NF-κB and MAPK activation in RAW 264.7 macrophages. Biochemical and Biophysical Research Communications. 2009;**381**(1):112-117. DOI: 10.1016/j.bbrc.2009.02.022

[108] Chang CL, Lin CS. Phytochemical composition, antioxidant activity, and neuroprotective effect of Terminalia chebula Retzius extracts. Evidence-Based Complementary and Alternative Medicine. 2012;**2012**:1-7. DOI: 10.1155/2012/125247

[109] Oleszek M, Oleszek W. Saponins in Food. Handbook of Dietary Phytochemicals. Singapore: Springer; 2020. DOI: 10.1007/978-981-13-1745-3_34-1

[110] Sparg S, Light ME, Van Staden J. Biological activities and distribution of plant saponins. Journal of Ethnopharmacology. 2004;**94**(2–3):219-243. DOI: 10.1016/j. jep.2004.05.016

[111] Ablise M, Leininger-Muller B, Dal Wong C, Siest G, Loppinet V, Visvikis S. Synthesis and in vitro antioxidant activity of glycyrrhetinic acid derivatives tested with the cytochrome P450/NADPH system. Chemical and pharmaceutical bulletin. 2004;**52**(12):1436-1439. DOI: 10.1248/cpb.52.1436

[112] Bouic PJ, Lamprecht JH. Plant sterols and sterolins: A review of their immune-modulating properties. Alternative Medicine Review. 1999;**4**(3):170-177

[113] Rasool M, Varalakshmi P. Immunomodulatory role of Withania somnifera root powder on experimental induced inflammation: An in vivo and in vitro study. Vascular Pharmacology. 2006;**44**(6):406-410. DOI: 10.1016/j.vph.2006.01.015

[114] Lee JH, Lee JY, Park JH, Jung HS, Kim JS, Kang SS, et al. Immunoregulatory activity by daucosterol, a β-sitosterol glycoside, induces protective Th1 immune response against disseminated candidiasis in mice. Vaccine. 2007;**25**(19):3834-3840. DOI: 10.1016/j.vaccine.2007.01.108

[115] Le CF, Kailaivasan TH, Chow SC, Abdullah Z, Ling SK, Fang CM. Phytosterols isolated from Clinacanthus nutans induce immunosuppressive activity in murine cells. International Immunopharmacology. 2017;**44**:203-210. DOI: 10.1016/j.intimp.2017.01.013

[116] Taheri Y, Suleria HA, Martins N, Sytar O, Beyatli A, Yeskaliyeva B, et al. Myricetin bioactive effects: Moving from preclinical evidence to potential clinical applications. BMC Complementary Medicine and Therapies. 2020;**20**(1):1-4. DOI: 10.1186/s12906-020-03033-z

[117] Chen CY, Peng WH, Tsai KD, Hsu SL. Luteolin suppresses inflammation-associated gene expression by blocking NF-κB and AP-1 activation pathway in mouse alveolar macrophages. Life Sciences.

2007;**81**(23–24):1602-1614. DOI: 10.1016/j.lfs.2007.09.028

[118] Mani R, Natesan V. Chrysin: Sources, beneficial pharmacological activities, and molecular mechanism of action. Phytochemistry. 2018;**145**: 187-196. DOI: 10.1016/j.phytochem. 2017.09.016

[119] Zhang X, Yang Y, Du L, Zhang W, Du G. Baicalein exerts anti-neuroinflammatory effects to protect against rotenone-induced brain injury in rats. International Immunopharmacology. 2017;**50**:38-47. DOI: 10.1016/j. intimp.2017.06.007

[120] Rosa SI, Rios-Santos F, Balogun SO, de Oliveira Martins DT. Vitexin reduces neutrophil migration to inflammatory focus by down-regulating pro-inflammatory mediators via inhibition of p38, ERK1/2 and JNK pathway. Phytomedicine. 2016;**23**(1):9-17. DOI: 10.1016/j.phymed.2015.11.003

[121] Feng L, Sun Y, Song P, Xu L, Wu X, Wu X, et al. Seselin ameliorates inflammation via targeting Jak2 to suppress the proinflammatory phenotype of macrophages. British Journal of Pharmacology. 2019;**176**(2):317-333. DOI: 10.1111/ bph.14521

[122] Liu F, Sun GQ, Gao HY, Li RS, Soromou LW, Chen N, et al. Angelicin regulates LPS-induced inflammation via inhibiting MAPK/NF-κB pathways. The Journal of Surgical Research. 2013; **185**(1):300-309. DOI: 10.1016/ j.jss.2013.05.083

[123] Stojanović-Radić Z, Pejčić M, Dimitrijević M, Aleksić A, Anil Kumar V, N, Salehi B, C Cho W, Sharifi-Rad J. Piperine-a major principle of black pepper: A review of its bioactivity and studies. Applied Sciences. 2019;**9**(20): 4270. DOI: 10.3390/app9204270

[124] Gao Y, Jiang W, Dong C, Li C, Fu X, Min L, et al. Anti-inflammatory effects of sophocarpine in LPS-induced RAW 264.7 cells via NF-κB and MAPKs signaling pathways. Toxicology In Vitro. 2012;**26**(1):1-6. DOI: 10.1016/j. tiv.2011.09.019

[125] Yue R, Jin G, Wei S, Huang H, Su L, Zhang C, et al. Immunoregulatory effect of Koumine on nonalcoholic fatty liver disease rats. Journal of Immunology Research. 2019;**2019**:8325102. DOI: 10.1155/2019/8325102

[126] Roy M, Liang L, Xiao X, Peng Y, Luo Y, Zhou W, et al. Lycorine downregulates HMGB1 to inhibit autophagy and enhances bortezomib activity in multiple myeloma. Theranostics. 2016;**6**(12):2209. DOI: 10.7150/thno.15584

[127] Checker R, Sandur SK, Sharma D, Patwardhan RS, Jayakumar S, Kohli V, et al. Potent anti-inflammatory activity of ursolic acid, a triterpenoid antioxidant, is mediated through suppression of NF-κB, AP-1 and NF-AT. PLoS One. 2012;**7**(2):e31318. DOI: 10.1371/journal.pone.0031318

[128] Ziaei S, Halaby R. Immunosuppressive, anti-inflammatory and anti-cancer properties of triptolide: A mini review. Avicenna J Phytomed. 2016;**6**(2):149

[129] Yang JH, Li B, Wu Q, Lv JG, Nie HY. Echinocystic acid inhibits RANKL-induced osteoclastogenesis by regulating NF-κB and ERK signaling pathways. Biochemical and Biophysical Research Communications. 2016;**477**(4): 673-677. DOI: 10.1016/j.bbrc.2016. 06.118

[130] Yu JS, Tseng CK, Lin CK, Hsu YC, Wu YH, Hsieh CL, et al. Celastrol inhibits dengue virus replication via up-regulating type I interferon and downstream interferon-stimulated responses. Antiviral Research. 2017;**137**: 49-57. DOI: 10.1016/j.antiviral.2016. 11.010

[131] Belapurkar P, Goyal P, Tiwari-Barua P. Immunomodulatory effects of triphala and its individual constituents: A review. Indian Journal of Pharmaceutical Sciences. 2014;**76**(6):467

[132] Gambari R, Borgatti M, Lampronti I, Fabbri E, Brognara E, Bianchi N, et al. Corilagin is a potent inhibitor of NF-kappaB activity and downregulates TNF-alpha induced expression of IL-8 gene in cystic fibrosis IB3-1 cells. International Immunopharmacology. 2012;**13**(3): 308-315. DOI: 10.1016/j. intimp.2012.04.010

[133] Garcia D, Leiro J, Delgado R, Sanmartin ML, Ubeira FM. Mangifera indica L. extract (Vimang) and mangiferin modulate mouse humoral immune responses. Phytotherapy Research. 2003;**17**(10):1182-1187. DOI: 10.1002/ptr.1338

[134] Zeng X, Guo F, Ouyang D. A review of the pharmacology and toxicology of aucubin. Fitoterapia. 2020;**140**:104443. DOI: 10.1016/j.fitote.2019.104443

[135] Bhaumik SK, Paul J, Naskar K, Karmakar S, De T. Asiaticoside induces tumour-necrosis-factor-α-mediated nitric oxide production to cure experimental visceral leishmaniasis caused by antimony-susceptible and-resistant Leishmania donovani strains. The Journal of Antimicrobial Chemotherapy. 2012;**67**(4):910-920. DOI: 10.1093/jac/dkr575

[136] Zhang YH, Isobe K, Nagase F, Lwin T, Kato M, Hamaguchi M, et al. Glycyrrhizin as a promoter of the late signal transduction for interleukin-2 production by splenic lymphocytes. Immunology. 1993;**79**(4):528

Chapter 10

Perspective Chapter: Utility of Injury Immunity Axis in Disease Phenotyping

Girish Kumthekar and Rajasekara Chakravarthi

Abstract

Organ injury is mediated by dysregulated inflammatory response of the host to invading organism or antigen. Dysregulated immune response can be more than or less than what is required to contain the organism or antigen. All disease states converge on inflammatory damage to tissue irrespective of what triggers the initial insult such as a transplanted organ, microbe, autoimmunity, and even a malignancy. Injury immunity axis can be used to phenotype a disease state to explaining its etiology, treatment options and possible disease trajectory. It will address the core issue of inflammation at cellular level guiding clinicians to tailor the treatment on case to case basis. This chapter brings immunity to center-stage in diagnosis and management of diseases due to various causes. This can be accomplished by phenotyping diseases across injury immunity axis to ascertain the status of immune system forefront. It is indeed a novel concept by which we look at different manifestations of a disease through a unique perspective. It is also an attempt to acknowledge the fact that immune system work-up and immune biomarkers need better representation in the list of investigations. The importance of immunological basis of diseases needs significant amount of research and robust data to translate this knowledge into the standard of care.

Keywords: injury immunity axis, immune function tests, translational immunology, immune biomarkers, disease phenotyping

1. Introduction

During the COVID-19 pandemic, we observed that it's hosts immune response to the invading SARS CoV2 that determines the clinical presentation and consequent outcomes. If immune response is appropriate, patients remain asymptomatic. If immune response is not appropriate, a cytokine storm becomes evident. And there are cases between these two extremes. It appears that COVID-19 is a disease of immune system.

This cannot be true for COVID-19 only. If we have a closer look, most if not all disease states are disorders of self-defense or immunity. Syndromes like sepsis, acute kidney injury can be explained by different immune responses by different hosts or by the same host at different time lines in the disease trajectory. These myriads of immune responses invariably present with various clinical scenarios. In fine, a single disease entity can present with different phenotypes depending on which type of immune response it has evoked in the host.

This manuscript is an attempt at having a look at immune system as a pivotal system in tissue injury. All disease states converge on inflammation and immune damages to tissue irrespective of what triggers the initial insult such as a transplanted organ, microbe, autoimmunity and even a malignancy. This manuscript is aimed at addressing the basic culprit and finding opportunities to address it for better treatment outcomes.

As of now we treat these immune system abnormalities without robust immune function tests. Tests like white cell count, differential count, procalcitonin, ferritin, hs-CRP, cytokine and chemokine assays are utilized as immune function tests. We need to understand disease states as abnormalities of immune system and disease phenotyping as per on one or more immune function test. Till we find better indices which may function as immune function tests, let's rely on whatever we have got and keep looking for better indicators of immune activation and suppression.

2. Basic concept

Organ injury is mediated by dysregulated immune/inflammatory response of the host to invading organism or antigen. Dysregulated immune response can be more than or less than what is required to contain the organism/antigen. Either way, dysregulated immune response causes tissue injury. But, the type of immune dysregulation will dictate the phenotype of injury. Medical or surgical management of tissue injury depends on the injury phenotype. In other words, we may need to typify the kind of immune dysregulation a patient has before initiating treatment.

Tissue injury can be prevented or resisted with optimal immune response. To complicate things further, immune response of the host during recovery determines if tissue recovers with regeneration or sclerosis (irreversible organ dysfunction). Hence, immune response of host to invading pathogen determines disease trajectory, treatment chosen and outcomes. With this background, we can have four different phenotype of tissue injury after an insult with invading organism/antigen. These four quadrants tell us if tissue injury is due to immune hyper-activation, immune deficiency or otherwise. Once we understand status of immune system, it's easier to manipulate the immune responses best suited for the host. This 2/2 model of injury immunity interplay can be exploited in a number of disease states where host has to deal with a foreign antigen. The foreign antigen may be a bacterium, virus, transplanted organ, malignant cell or even a self-antigen to which tolerance is lost (**Figure 1**).

2.1 COVID-19

In response to SARS-CoV-2 infection, both innate and adaptive immune systems are involved. SARS-Co-V applies several mechanisms to overcome the immune response. First, it inhibits the rapid expression of interferon type 1 (IFN-1). Moreover, SARS-CoV-2 interferes with IFN-1 signaling through inhibition of STAT-1 phosphorylation. The second defensive mechanism of SARS-CoV-2 is immune exhaustion through exaggerated and prolonged IFN-1 production by plasmacytoid dendritic cells (pDCs). This process leads to the influx of activated neutrophils and inflammatory monocytes/macrophages, that in turn, results in lung immunopathology (e.g. acute respiratory distress syndrome). Finally, it results so-called "cytokine storm" further weakens the immune system through IFN-1 mediated T cell apoptosis [1, 2].

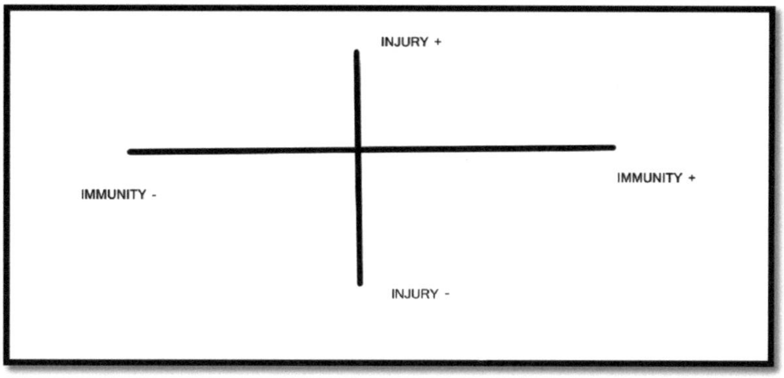

Figure 1.
Injury immunity axis is plotted as a 2/2 model. Right upper quadrant has hyperactive immune response causing tissue injury. Right lower quadrant shows optimal immune response with no tissue injury, hence host stays protected. Left upper quadrant shows deficient immune response that leads to tissue damage as direct cytotoxicity of invading organism/antigens. Left lower quadrant belongs to an unexposed host or an exposed host deficient in receptors for the invading organism to adhere and internalize.

Immune classification of patients with SARS-CoV-2 was performed by using the tools suggested for bacterial sepsis, i.e. ferritin more than 4420 ng/ml for MAS, and HLA-DR molecules on CD14 monocytes lower than 5,000 in the absence of elevated ferritin, for the immune dysregulation phenotype. It was found that contrary to the patients with bacterial community acquired pneumonitis and severe respiratory failure (SRF), all patients with SRF and SARS-CoV-2 had either immune dysregulation or macrophage activation syndrome (MAS) [3, 4].

It looks like COVID-19 is a disorder of immune system and host immune dysregulation determines disease phenotype from asymptomatic carrier state to florid pneumonia. Different disease phenotypes are not only important for diagnosis but gives us a chance to modify treatment on case to case basis. It is more than obvious that one disease phenotype may not get appropriate benefits from treatment aimed at other phenotypes (**Figure 2**).

2.2 Sepsis

Immune responses of critically ill patients with sepsis is classified into three patterns: macrophage-activation syndrome (MAS), sepsis-induced immune-paralysis characterized by low expression of the human leukocyte antigen D (HLA-DR) on CD14 monocytes and an intermediate functional state of the immune system lacking obvious dysregulation [3].

In contrast with sepsis-1, in sepsis-3, the “systemic inflammatory response” is replaced with “dysregulated host response”, and SIRS was changed to SOFA. The dysregulation of host responses is a complicated process and includes inflammation, the neuroendocrine response, coagulation, and metabolic responses. In fact, the neuroendocrine response and coagulation are closely linked to inflammation [5].

Hyper-inflammation-induced organ failure is thought to be the most common cause of death during the first days of sepsis. In the chronic phase of sepsis, persistent inflammation, immunosuppression, and catabolic syndrome (PICS) becomes the main cause of secondary ICU-acquired infections and long-term mortality. However, it is difficult to distinguish hyper-inflammation and immunosuppression in patients with sepsis. Therefore, the timing and course of anti-inflammatory treatments also

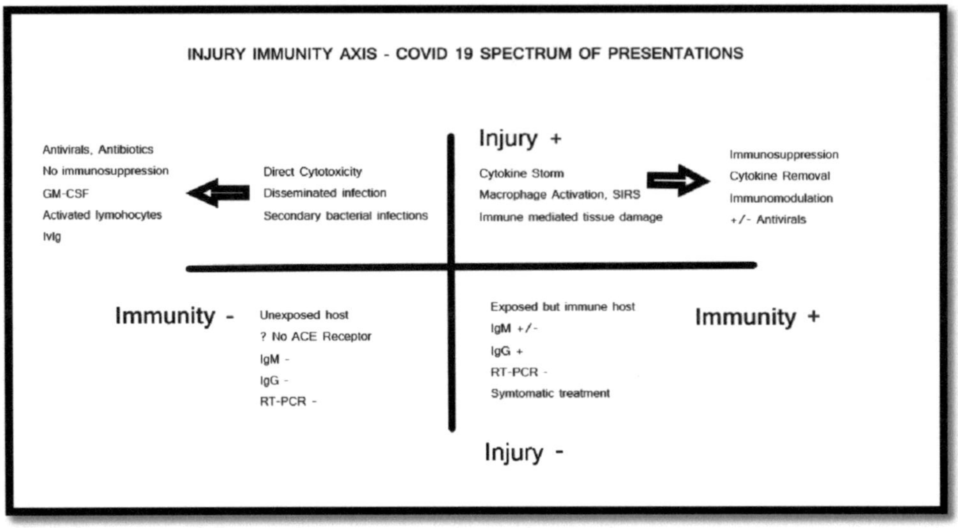

Figure 2.
This diagram shows clinical spectrum of COVID 19 presentations integrated on injury immunity axis. Right upper quadrant indicates a patient with cytokine storm with hyperactive immune system causing tissue injury which is managed with immune suppression and extracorporeal cytokine/chemokine removal. Right lower quadrant shows optimal host immune response which is protective with appropriate antibody production and preventing tissue injury. Left upper quadrant belongs to a host unable to mount immune response (hypoactive immune response), hence shows extensive viral replication, direct cytotoxicity requiring immune augmentation, antiviral and anti-bacterial. Left lower quadrant belongs to a person not exposed to SARSCoV2 or who is theoretically deprived of receptor for viral attachment and internalization (ACE II). In either way, such a person remains protected with no tissue injury.

require discussion. Assessing the inflammatory and immune status of sepsis calls for a precise stage-dependent therapy [6].

Current sepsis observations suggest that multiple organ failure occurs even in the context of preserved cell morphology. In addition, organ dysfunction is often reversible, even in organs that regenerate poorly (heart, lung, central nervous system, kidneys). Therefore, it is apparent that sepsis-induced organ dysfunction occurs primarily though cellular and molecular dysregulation, as opposed to gross tissue damage. By this principle, immune dysfunction in sepsis is also associated with molecular alterations that alter cellular phenotype and function.

Moreover, not only acquired but innate immune system is activated in sepsis and contributes to tissue damage. There is well established evidence that activation of the complement system is often linked to activation of both the clotting and the fibrinolytic systems. Development of neutralizing C5a antibodies in murine models dramatically attenuated the intensity of sepsis, including greatly improved 7th day survival, reduced levels of plasma cytokines, and decreased multiple organ failure. The vast majority of patients who die from sepsis have ongoing infections, suggesting that defects in innate immunity in general and neutrophil-mediated bacterial clearance in particular, could serve as potential therapeutic targets to regulate neutrophil apoptosis, production, maturation and function [7]. Given the profound immunosuppression induced by depletion of immune cells that occurs during sepsis, the ability to sequentially follow the uncontrolled lymphocyte apoptosis as a means to evaluate the efficacy of immune adjuvant therapies provides promising novel therapeutic opportunities. Furthermore, an increasing number of immune-adjuvant therapies to prevent

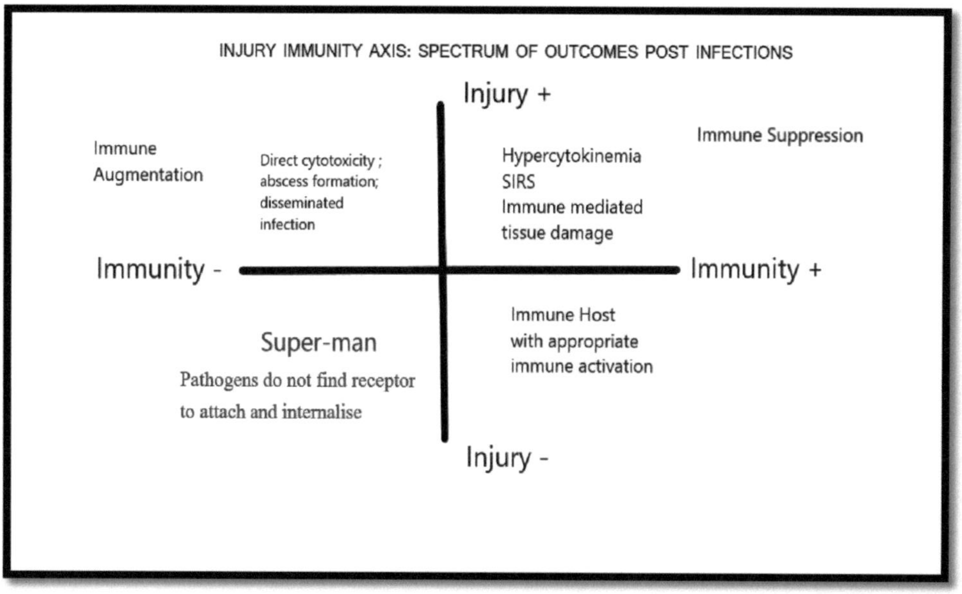

Figure 3.
It shows sepsis phenotypes integrated on the injury immunity axis. Right upper quadrant shows hyperactive immune response seen in SIRS, hypercytokinemia and macrophage activation which is dealt with immunosuppression and/or extracorporeal cytokines removal. Right lower quadrant belongs to appropriate immune activation hence no tissue injury. Left upper quadrant is sepsis induce immune deficiency (so called immune paresis/paralysis). Tissue damage in patients belonging to this quadrant is due to direct cytotoxicity, disseminated infection and needs immune augmentation. Left lower quadrant is a curious case of someone tolerant to invading pathogen/antigen owing to lack of exposure to antigen or theoretically absent pathogen specific receptors.

sepsis-induced immune paralysis have been identified as apoptosis dependent. IL-7 and anti-PD-L1 have been found to have potent effects to prevent lymphocyte apoptosis [8].

In fine, unless we know the phenotype of sepsis manifestation, choosing appropriate therapy for the immune system dysfunction is difficult to practice (**Figure 3**).

2.3 Organ transplantation

Organ transplantation is threatened by the possibility of graft loss due to rejection. Rejection is the immune response by the recipient immune system that injures graft parenchyma. Based on the chronology, severity and insidiousness, rejections are popularly known as hyperacute, acute or chronic. But, end result of all rejections is graft loss unless treated with appropriate immunosuppression. Appropriateness of immunosuppression is gauged by graft organ function and not by immune function tests. We know that graft function decline is a late feature compared to immune activation and invasion of graft tissue. But, management of allograft dysfunction is not yet based on immune function tests (immunometer).

The survival of ABO incompatible-transplanted (ABOi) organs in coexistence with anti-allograft antibodies and complement which originally results in graft rejection was described as accommodation. Is this successful engraftment of ABOi allografts (accommodation) a certain level or type of allograft tolerance, or does it just reflect some other biological condition of allografts? Thus, the mechanism investigation of accommodation and tolerance could be significant for conquering humoral barriers to transplantation

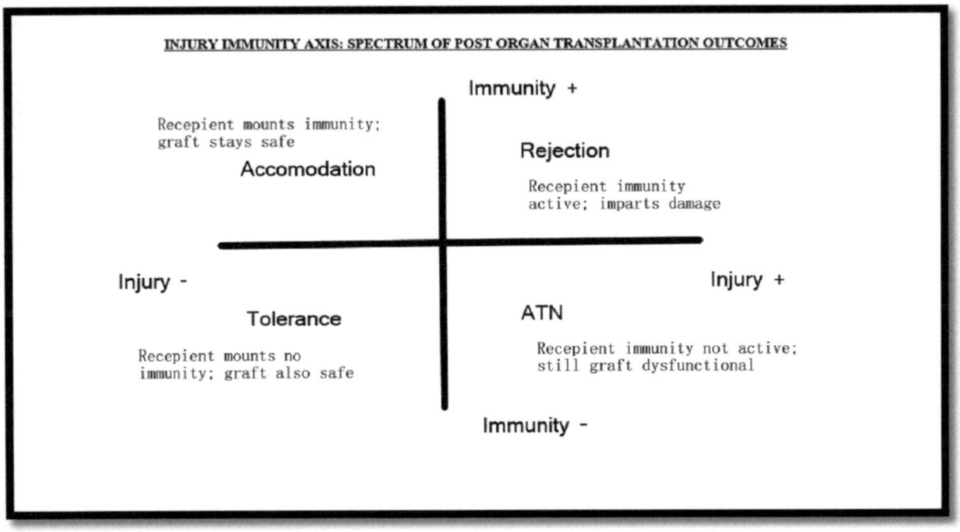

Figure 4.
It shows sequelae to organ transplantation with 2/2 injury immunity axis model. Right upper quadrant shows graft injury which is immune mediated (rejection) which needs immunosuppression as therapy. Right lower quadrant highlights graft injury due to causes other than rejection (drugs, infections, ischaemia) where immune suppression may or may not be excessive. Left upper quadrant show no injury to graft despite host mounting immune response. This would arise as the graft is able to withstand the immune activation (accommodation). Left lower quadrant shows inactive host immunity (spontaneous or induced) and consequent no injury to grafted tissue (tolerance). But it should be apparent that these are not water tight compartments and patient can move from one to another quadrant. Clinicians may aim to reach left upper quadrant phenotype for their patients with all possible therapies.

and promoting long-term survival of allografts. Tolerance is a state of the immune system unresponsiveness to substances or tissue that are capable of eliciting an immune response in a given organism, in contrast with traditional immune-mediated elimination of foreign antigens. Accommodation is a unique immunologic condition that is different from immune tolerance. It is defined operationally as a state in which the transplanted organ works normally under the existence of antibodies in the recipient specifically targeting the allograft [9]. As of now we do not have any specific tests that might tell us occurrence and sustenance of either accommodation or tolerance.

Another issue that crops up in graft dysfunction is infections, drug toxicities and ischemia. In the absence of immune function tests (read immunometer), differentiating drug toxicities, viral infections (BKVN, CMV) from rejection becomes difficult. Needless to mention that these entities represent excessive immunosuppression on most of the occasions.

As organ transplantations and its sequelae are determined by immune response of recipient, we need to base treatments on immune function tests rather than wait till graft dysfunction becomes evident and at times insurmountable. Management of allograft recipient can be tailor-made if the patient is allotted to any of the injury immunity axis quadrants (diagram 4). This may give us an opportunity to alter immunosuppression before the graft shows decline in function or recipient shows adverse effects of immunosuppression (**Figure 4**).

2.4 Acute kidney injury

Acute kidney injury is a clinical syndrome resulting from multiple etiologies. Presently, KDIGO classification for defining AKI is based on functional parameters of rising serum creatinine and drop in urine output. With evolution in understanding of AKI pathogenesis, it is well understood that damage to kidney precedes decline in function. Hence, there are efforts ongoing on how to integrate functional markers and damage markers in defining AKI. As it holds true for other organ systems, kidney damage is mediated and sustained by inappropriate inflammation and immune activation. This immune mediated damage is equally sustained by cortex and medullary compartments. Hence, if we intend to reverse/modify or transform the immune activation, treatment should be based on immune mediated damage phenotype. In other words, immune dysfunction preceded damage to kidney and consequent AKI.

Let's try to come up with a model where immune biomarkers (cause), damage biomarkers (effect) and functional biomarkers (presentation) throw light on AKI phenotype which will guide management of this syndrome easier.

Inflammation is a complex biologic response that is essential for eliminating microbial pathogens and repairing tissue after injury. AKI associates with intrarenal and systemic inflammation; thus, improved understanding of the cellular and molecular mechanisms underlying the inflammatory response has high potential for identifying effective therapies to prevent or ameliorate AKI. Coupled to this is the emerging concept that mechanisms of intrarenal inflammation during AKI also exert potentially harmful effects on distant organs and tissues through release of soluble mediators or re-entry of activated leukocytes into the bloodstream [10].

Research focused on identifying and measuring inflammatory phases in the setting of human AKI is required to target inflammation modifying therapies and identify optimal times to start and stop such therapies. Additionally, great emphasis should be given to comparative analysis regarding the nature and kinetics of inflammatory response in AKI (**Figure 5**) [11].

2.5 Autoimmune disorders

Tolerance is the failure of the immune system to respond to an epitope in an aggressive way. Most self-tolerance results from the deliberate inactivation or destruction of lymphocytes bearing BCRs or TCRs (B cell and T cell receptors) that recognize and bind self-epitopes. Inactivation or destruction may occur during early development (central tolerance) or may be imposed on lymphocytes in the periphery (peripheral tolerance).

Under normal circumstances, autoreactive cells in the body are not activated by contact with self-molecules. Unless they are interacting with APCs, they are not also receiving cytokine signals necessary for activation. However, in inflammatory sites, local cytokine levels may be sufficient to activate auto reactive T cells when they are binding to self-epitopes on non-APCs [12].

Thus, in a patient with autoimmune disorder, two mutually independent immune systems are at work. Autoimmune system is aberrant, unchecked and harmful to the host, hence needs to be suppressed. Alloimmunity is trigger operated, responds to feedbacks and protective to host. But, when immunosuppression is started for a person with autoimmune disorder, none of the immune function tests are used to guide standard of care for various reasons such as no confirmed benefits, lack of cost effectiveness and inability to transform into clinical outcomes. In these situations, immune system aberrations are dealt, not according to immune status of host (which may be described by an immunometer) but according to target organ function. This leads to excessive or inadequate immunosuppression. In other words, it's preferable to

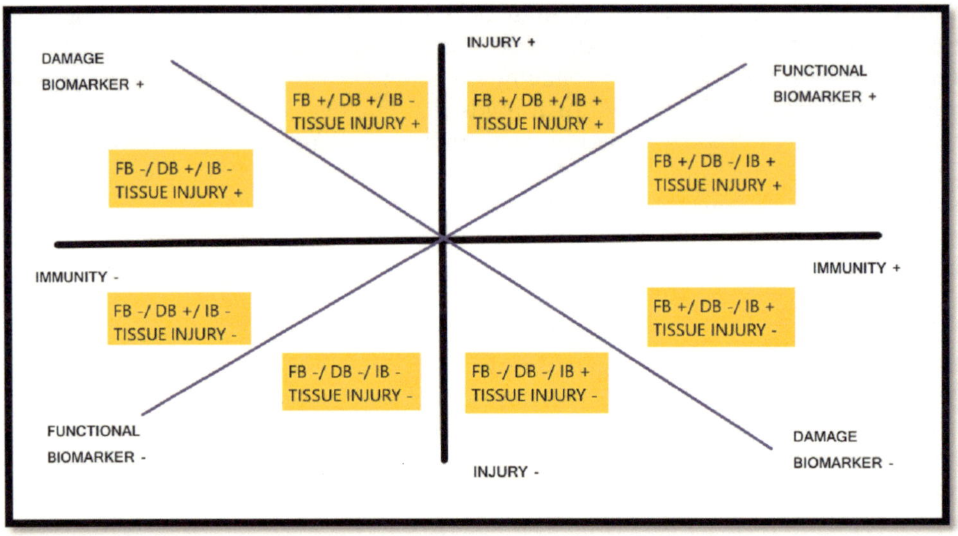

Figure 5.
AKI phenotyping is described using three sets of biomarkers 1. functional 2. damage 3. immune biomarkers. This gives us eight different AKI phenotypes and their possible management options. 2/2 model of functional and damage biomarkers is superimposed on injury immunity axis to further understand the chain of events in AKI syndrome. Functional biomarker positivity indicates CKD or advanced AKI. Damage biomarker positivity indicates early AKI which may be immune mediated (IB positive) or non-immune mediated (IB negative). Theoretically, let's presume IB appears earlier than DB and disappears late in the disease trajectory of AKI. FB: functional biomarker; DB: damage biomarker; IB: immune biomarker.

suppress autoimmunity selectively and preserve alloimmunity with the help of best possible immune function tests.

Interleukin-6 (IL 6) is elevated in the sera of SLE patients and is considered a sensitive marker of disease activity and nephritic flares. It is a potent stimulator of the differentiation and activation of lymphoid and myeloid cells and the production of acute phase proteins within the liver [13, 14]. But as we know, none of the management guidelines on autoimmune disorders recommend immune function tests as a part of standard of care (**Figure 6**).

3. Quest for immunometer

Immunometer, as we may call it, is a number that tells us status of immune system (innate immunity, acquired immunity and autoimmunity). It is essentially at a conceptual level and may be a vital component of immune system if proves validated. Multiple molecules are candidate for immunometer but none stands proven as of now. It could be because of lack of robust data on tissue injury phenotyping or traditional reliance on target organ function parameters to access immune system status and obvious complexity of immune system itself. We observed that in above mentioned 2/2 models for different disease states injury immunity axis gave us desired phenotypes. Injury immunity axis can be exploited to find different disease phenotypes explaining etiology, treatment options and possible disease trajectory. It will address the core issue of inflammation at cellular level guiding clinicians to tailor treatment case to case

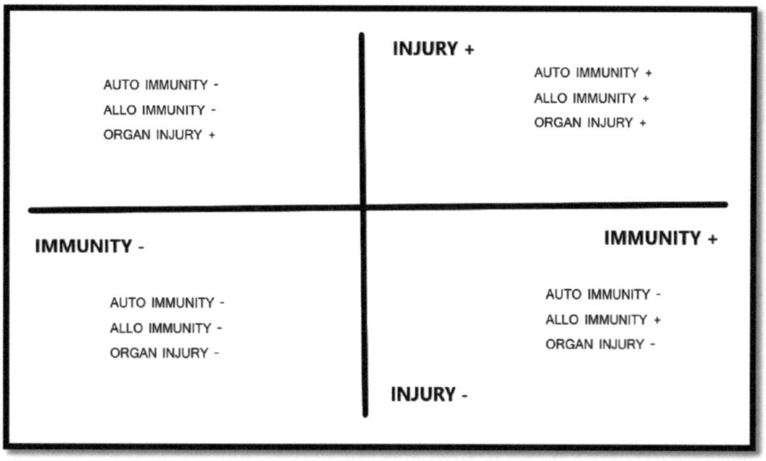

Figure 6.
It shows injury immunity axis for autoimmune disorders. There are two different immune systems at work 1. Alloimmune and 2. Autoimmune. Right upper quadrant shows unchecked hyperactive autoimmunity causing tissue damage. Post immunosuppressive therapy patient may migrate to right lower quadrant with reversal of tissue injury with preserved alloimmune response. If immune suppression is continued or escalated, patient reaches left lower quadrant with suppressed alloimmune response making host susceptible for infections, drug toxicities and consequent tissue injury (left upper quadrant). Clinicians may prefer to hold patients in right lower quadrant as long as possible (state of disease remission).

basis and avoid unnecessary interventions. The immunometer can consist of three types of biomarkers describing current status of immune system. First type of biomarkers can describe dampened immune response (immune paralysis) and the prototype biomarker can be IL-10. Second type can be a set of biomarkers exhibiting excessive immune response (immune hyperactivation) and the prototype biomarker can be interleukin 6/12, TNF alpha. The third type can describe immune protection (appropriate immune response) to the host and the prototype biomarker can be IgG against the particular antigen or transforming growth factor beta. Although produced by a wide variety of cell types, macrophages and T lymphocytes (T cells) are the primary producers of cytokines, which may have predominantly pro-inflammatory (inflammation-promoting; IL1α, IL1β, IL2, IL6, IL8, IL12, TNFα, IFNγ) or anti-inflammatory (inflammation-suppressive; IL4, IL5, IL10, TGFβ) abilities [15, 16]. The relative contribution of various biomarkers can be explained with the help of immunometer. This is a novel concept and needs further research. The immunometer concept will stimulate further research and pave way for the discovery of novel biomarkers.

To explain the concept, let us see an example of how important immune function testing can be. 53-years-old male patient presented with complaints of fever, dysuria and malaise for 1 week. On evaluation, he was diagnosed as bilateral pyelonephritis with right sided emphysematous pyelonephritis. As a result, patient developed urosepsis, acute kidney injury (AKI) on chronic kidney disease (CKD). Due to severe azotaemia and acidosis, patient required dialytic support (SLED). Though there was partial improvement initially, he again had reappearance of high-grade intermittent fever. This warranted for a right sided PCNL and drainage of perinephric collection. To look for other causes of non-responding sepsis, serum ferritin and interleukin 6 (IL-6) were tested. Both of these biomarkers were significantly elevated.

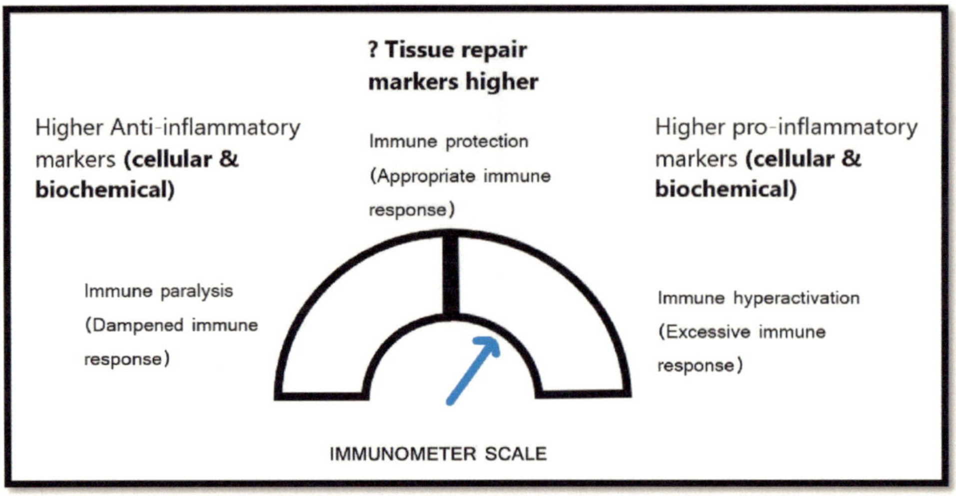

Figure 7.
It shows a model of immunometer. It is akin to a galvanometer. On left hand side of center, it shows extent of immune suppression (dampened immune response) and on right hand side, it shows immune hyperactivation (excessive immune response or a cytokine storm). The central part is an appropriate immune response of the host which is either protective or reparative. The knowledge of the type of immune response a particular patient exhibits on the immunometer scale can be helpful in diagnosis of disease phenotype, choosing treatment modality and prognostication as well. Immunometer concept can be applied to multiple biomarkers at a time that may change the outlook of multiple clinical presentations of one particular disease.

(Ferritin 4532 ng/ml and IL-6 531 pg/ml on 1.8.19). Bone marrow aspiration and biopsy was done which revealed acquired hemophagocytosis with no evidence of myelodysplasia, granulomas or infiltration. Patient was started with hydrocortisone 200 mg/day. Gradually fever subsided and hemodynamic improved. As patient showed adequate respiratory attempts and acceptable oxygenation, he was extubated [17].

Seemingly, treating immunological disorders and sepsis is seen antagonistic but might be mutually complementary as is seen in this particular case with acquired HLH (hemophagocytic lymphohistiocytosis) where patient responded to immune suppression with corticosteroid. Hence, biomarkers like ferritin, IL 6 can constitute the immunometer and can help diagnose underlying immune dysfunction.

Moreover, antibiotic resistance and scarcity of new antibiotics are two more reasons to search for immune dysregulation in sepsis. As many of these cases could be immune dysregulation and not infection per se (**Figure** 7).

4. Candidate molecules/cells for constructing an immunometer

Multiple of molecules and cells contribute to immune functions. Many of these are specific for a particular arm of immune system. We tried to compile most of these molecules/cells for finding out appropriate candidate/s for describing immune system as immunometer (**Table 1**).

Innate immune system		**Acquired immune system**	
Cells	**Substances**	**Cells**	**Substances**
aAPC, aDC, Macrophages (M1, M2),	CH 50, C3, C4, Perforins, Granzymes, soluble fas ligands	CD14, Monocyte HLA DR, Eosinophis, actiavated T cELLS, Treg, Breg, TH17, CD 4+, cd 8+, neutrophil to lymphocyes ratio (NLR), platelet to lymphocyte ratio (PLR)	IL 1, IL6, IL 10, sTNF receptor, TNF, TGF-b, chemokines, resolvin, protectin D1, Heat shock proteins
Autoimmunine system			
Post organ transplantation (alloantigen)		**Native tissue antigen (autoantigen)**	
Cells	**Substances**	**Cells**	**Substances**
Immunophenotyping Organ biopsy,	Donor specific antibodies (HLA/NON-HLA), urine perforins, CD antigens (14, 3, 4)	TH1/TH2 ratio, CD 4+/25+ cells, neutrophil to lymphocyes ratio (NLR), platelet to lymphocyte ratio (PLR)	ANA, ANCA, DS-DNA, anti GBM antibody,

The Table shows three types of immune systems at work. The innate, acquired and auto immune systems are closely related and affect each other in many ways. To understand their respective contribution to a particular disease state, we need distinct biomarkers. These biomarkers will constitute the hypothetical immunometer helping clinicians decide on targeted treatment to modify the particular immune dysregulation. A lot of work and data is needed before we bring the immunometer hypothesis to reality.

Table 1.
Candidate molecules/cells for constructing an immunometer.

5. Conclusions

This article is an attempt to understand multiple pathways for a particular disease state to manifest in different phenotypes. Each particular phenotype having a distinct immune function abnormality requiring a precise targeted treatment. Unless we understand the immune interplay among different phenotypes of a single disease, it may not be possible to address the underlying immune function abnormality for achieving better outcomes.

Author details

Girish Kumthekar[1,2*] and Rajasekara Chakravarthi[1,3]

1 RENOWN Clinical Services, Hyderabad, India

2 Symbiosis Medical College for Women and Symbiosis University Hospital and Research Centre, Symbiosis International University (SIU), Pune, India

3 Star Hospitals, Banjara Hills, Hyderabad, India

*Address all correspondence to: drgirishkumthekar@gmail.com

References

[1] Prompetchara E, Ketloy C, Palaga T. Immune responses in COVID-19 and potential vaccines: Lessons learned from SARS and MERS epidemic. Asian Pacific Journal of Allergy and Immunology. 2020;**38**(1):1-9. DOI: 10.12932/AP-200220-0772. PMID: 32105090

[2] Channappanavar R et al. Dysregulated Type I interferon and inflammatory monocyte-macrophage responses cause lethal pneumonia in SARS-CoV-infected mice. Cell Host & Microbe. 2016;**19**(2):181-193

[3] Qin C, Zhou L, Hu Z, Zhang S, Yang S, Tao Y, et al. Dysregulation of Immune Response in Patients With Coronavirus 2019 (COVID-19) in Wuhan, China. Clinical Infectious Diseases. 2020 Jul 28;**71**(15):762-768. DOI: 10.1093/cid/ciaa248. PMID: 32161940; PMCID: PMC7108125

[4] Complex immune dysregulation in COVID-19 patients with severe respiratory failure. Cell Host & Microbe. 2020;**27**:1-9

[5] Ding R, Meng Y, Ma X. The central role of the inflammatory response in understanding the heterogeneity of sepsis-3. BioMed Research International. 2018;**2018**:10. DOI: 10.1155/2018/5086516 [Article ID: 5086516]

[6] Bermejo-Martin JF, Andaluz-Ojeda D, Almansa R, et al. Defining immunological dysfunction in sepsis: A requisite tool for precision medicine. Infection. 2016;**72**(5):525-536

[7] Boomer JS et al. Immunosuppression in patients who die of sepsis and multiple organ failure. Journal of the American Medical Association. 2011;**306**:2594-2605

[8] Cao C et al. Cell death and disease. Official Journal of the Cell Death Differentiation Association. 2019;**10**:782

[9] Wang J, Ruiming R. Mechanism Development of Accommodation and Tolerance in Transplant. In: ABO-incompatible Organ Transplantation. DOI: 10.1007/978-981-13-3399-6_15. Springer. ISBN: 978-981-13-3398-9

[10] Ratliff BB, Rabadi MM, Vasko R, Yasuda K, Goligorsky MS. Messengers without borders: Mediators of systemic inflammatory response in AKI. Journal of American Society of Nephrology. 2013;**24**:529-536

[11] Rabb H, Griffin MD, McKay DB, Swaminathan S, Pickkers P, Rosner MH, et al. Inflammation in AKI: Current understanding, key questions, and knowledge gaps; The acute dialysis quality initiative consensus XIII Work Group. Journal of American Society of Nephrology. 2016;**27**:371-379

[12] Autoimmunity: Lippincott Textbook of Immunology. 2nd edition. Lippincott Williams & Wilkins; 2012. p. 376. ISBN-10: 9781451109375. ISBN-13: 978-1451109375

[13] Dima A et al. Clinical significance of serum and urinary interleukin-6 in systemic lupus erythematosus patients. Egyptian Rheumatology. 2017;**39**(1):1-6

[14] Hahn B et al. American college of rheumatology guidelines for screening, treatment, and management of lupus nephritis. Arthritis Care and Research. 2012;**64**(6):797-808

[15] Dinarello CA. Proinflammatory cytokines. Chest. 2000;**118**:503-508

[16] Zhou X, Fragala MS, McElhaney JE, Kuchel GA. Conceptual and methodological issues relevant to cytokine and inflammatory marker measurements in clinical research. Current Opinion in Clinical Nutrition and Metabolic Care. 2010;**13**: 541-547

[17] Kota M, Kumthekar GV, Rajasekara Chakravarthi M, Guttikonda J, Srinivas G. Acquired Hemophagocytic Lymphohystiocytosis in A Patient with Emphysematous Pyelonephritis and Severe Sepsis: A case report. Nephrology Dialysis Transplantation. 2020;**302** (3142P). DOI: 10.1093/ndt/gfaa142. P0333

Chapter 11

Perspective Chapter: Drug-Induced Severe Cutaneous Adverse Reactions, Diagnostics and Management

Miteshkumar Rajaram Maurya, Renuka Munshi, Sachin Bhausaheb Zambare and Sanket Thakur

Abstract

Severe cutaneous Adverse Reactions (SCAR) are rare drug hypersensitivity reactions but can be life-threatening if not appropriately and timely managed. Many research studies have shed light on its pathomechanism and triggers that have helped us better understand SCAR. The presence of viral fever and genetics such as HLA genotype with certain drugs have been associated with the occurrence of SCAR. However, the basis of interaction of these causative agents needs further evaluation to understand the predisposition to the reaction occurrence. The different spectrum of SCAR needs to be clinically diagnosed appropriately which includes Drug Reactions with Eosinophilia and Systemic Symptoms (DRESS), Steven Johnson Syndrome/Toxic Epidermal Necrolysis (SJS/TEN), Acute Generalized Exanthematous Pustulosis (AGEP), and generalized bullous fixed drug eruptions (GBFDE). However, due to the rare occurrence of this reaction, there is not sufficient evidence for the best treatment for patients suffering from SCAR. Our review provides detailed information about the disease type, manifestation, pathophysiology, diagnostics, and current treatment aspects of SCAR.

Keywords: SCAR, adverse drug reaction, cutaneous eruptions, Steven Johnson syndrome, toxic epidermal necrolysis, acute generalized exanthematous pustulosis, clinical pharmacologists

1. Introduction

Since time immemorial, medications come with some benefits and risks. However the intention of treating patients to cure their ailments remains the utmost priority of all the physicians. Thus stands true, the famous dictum by Hippocrates (460–370 BC) that states "Primum non-nocere" which means first of all be sure you do no harm and benefit come next. The reality however is that at times, risk and benefit cannot be separated out so what we follow today is as long as the benefit outweighs the risk, we are ready to take risks in every walk of life similar to what we do for medicines.

Severe Cutaneous Adverse Reactions (SCARs) are drug associated hypersensitivity reactions that involves skin and mucous membranes of various body orifices such as eyes, ears, the inner surface of the nose, buccal mucosa, and lips and may involve damage to internal organs usually mediated by drug specific T lymphocytes [1, 2]. The phrase "Skin is Like an Ocean's Surface Which Tells Deep Stories If You Watch Carefully" goes very well for SCARs which is a multi-spectrum disease due to its variable manifestations. SCAR has been classified into five main types- Drug Reaction with Eosinophilia and Systemic Symptoms (DRESS)/Drug-Induced Hypersensitivity Syndrome (DIHS), Steven Johnson Syndrome (SJS), Toxic Epidermal Necrolysis (TEN), SJS/TEN overlap syndrome and Acute Generalized Exanthematous Pustulosis (AGEP) [3]. Severe Cutaneous Adverse Reactions (SCAR) terminology was proposed for such conditions, as they were (a) severe, (b) unpredictable, and (c) drug-induced [4]. The pathophysiology remains almost similar in all these five types of severe cutaneous adverse reactions. However, immunological trigger due to viral, drug, and gene interaction is still unknown that requires intensive research. Even the predictability, diagnosis, and treatment remains challenging and uncertain for most dermatologist, immunologists, and clinical pharmacologists.

1.1 Classification of severe cutaneous adverse drug reactions (SCAR) provided by the World Allergy Organization (2014)

1. Steven Johnson Syndrome

2. Toxic Epidermal Necrolysis

3. Steven Johnson Syndrome/Toxic Epidermal Necrolysis Overlap Syndrome [5].

4. Drug-induced Hypersensitivity Syndrome (DIHS) or Drug reaction with Eosinophilia and Systemic Symptoms (DRESS)

5. Acute Generalized Exanthematous Pustulosis (AGEP)

1.2 Coombs & Gell's classification of hypersensitivity skin reactions

Table 1 is summarized below.

Hypersensitivity reaction	Predominant inflammatory cells	Clinical Conditions
Type I	Immediate, IgE-mediated mast cell activation	Anaphylactic reaction to bee sting, antibiotic like penicillin, latex allergy.
Type II	Antibodies, cytotoxic, IgG-/ IgM-mediated, complement	Hemolytic reactions Good pasture Syndrome Hyper acute graft rejections
Type III	mediated by immune complexes and IgG/IgM, complement	Hypersensitivity Pneumonitis (HP) Systemic Lupus Erythematosus (SLE) Polyarteritis Nodosa (PAN) Serum sickness
Type IV	Delayed-type hypersensitivity reactions mediated by T-helper and T-cytotoxic cells	Chronic Allograft rejection Purified Protein Derivative (PPD) test Latex/Nickel/ Poison ivy allergy

Hypersensitivity reaction	Predominant inflammatory cells	Clinical Conditions
Grading of Type IV hypersensitivity skin reactions by Gell and Coombs		
Type IVa	Monocytes	Drug-induced maculopapular exanthems
Type IVb	Eosinophils	Drug Reaction with Eosinophilic Systemic Symptoms (DRESS)
Type IVc	Cytotoxicity of drug specific T cells	Steven Johnson Syndrome/Toxic Epidermal Necrolysis
Type IVd	Neutrophils	Acute Generalized Exanthematous Pustulosis (AGEP)

Table 1.
Coombs and Gell classification of hypersensitivity cutaneous reactions.

2. Epidemiology of severe cutaneous adverse reactions (SCARs)

The skin is the most common and easily visible body part that manifest symptom of adverse drug reactions [6]. Adverse Drug Reactions (ADRs) contribute up to 7% of the hospital admissions of which cutaneous ADRs alone contribute to 2–3% of the overall hospitalizations [7, 8]. Few of these cutaneous adverse drug reactions have disabling sequelae or can be life-threatening. The incidence of fatalities among inpatients due to systemic and cutaneous adverse drug reactions reported ranges between 0.1% to 0.3%. However, the mortality rates for SJS/TEN are approximately 5–10%, 30–50% in TEN, 10% in DRESS, and the common leading culprit drugs associated with SCARs, around the world are antibiotics, anti-epileptics, allopurinol, Non-steroidal anti-inflammatory drugs (NSAIDS), and antiretrovirals [9–11]. Incidence rates of SJS/TEN range from 1.4–12.7/million person-years in different studies [12–14]. A Nationwide Korean Study of Severe Cutaneous Adverse Reactions by Kang DY et al. based on the Multicenter Registry revealed total of 745 SCAR cases (384 SJS/TEN cases and 361 DRESS cases) due to 149 drugs. The main causative drugs suspected were allopurinol (14.0%), carbamazepine (9.5%), vancomycin (4.7%), and anti-tubercular agents (6.3%). Carbonic anhydrase inhibitors (100%), nonsteroidal anti-inflammatory drugs (84%), and acetaminophen (83%) were common offending drugs for SJS/TEN whereas dapsone (100%), antituberculous agents (81%), and glycopeptide antibacterials (78%) were associated with DRESS. The overall mortality rate reported in this study due to SCARs cases was 6.6% (SJS/TEN-8.9% and DRESS-4.2%) [15]. The phase 1 of regiSCAR project by the European community (last updated on October 25, 2014), includes potential 1889 (69.39%) cases of SJS/TEN/GBFDE/EEMM, 364 (13.37%) cases of AGEP, 469 (17.22%) cases of DIHS/DRESS. Of these suspected cases, the definite or probable cases of SJS/TEN- 1232 (65.21%), EEMM- 251 (13.28%), GBFDE- 5 (0.26%), AGEP- 228 (62.63%) and HSS/DRESS- 281 (59.91%) [16].

3. Steven Johnson syndrome/toxic epidermal necrolysis (SJS-TEN) overlap syndrome

3.1 History of SJS and TEN

Steven Johnson Syndrome was first described in 1922 by American pediatrician named Albert Mason Stevens (1884–1945) and Frank Chambliss Johnson (1894–1934) who reported two cases of boys in New York City, United States, aged 7 years and 8 years presenting with an extraordinary generalized skin eruption with continued fever, inflamed buccal mucosa, and severe purulent conjunctivitis [17]. Both these cases were misdiagnosed by primary care physicians as a case of "hemorrhagic measles" or "black measles". Similarly, the first description of Toxic Epidermal Necrolysis (TEN) also called Lyell syndrome was given by Scottish dermatologist Dr. Alan Lyell (1917–2007) in 1956 in four patients. Initially, these were considered as the toxic eruption that resembled severe burn or scalding of skin associated with erythematous plaques and widespread areas of epidermal detachment and was referred

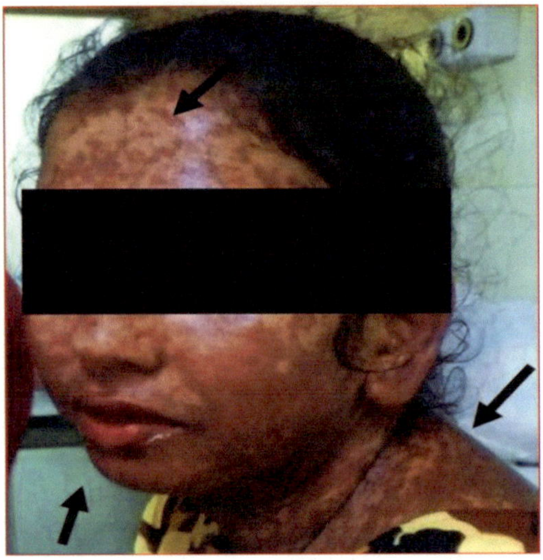

Figure 1.
Epidermal involvement of the face and neck.

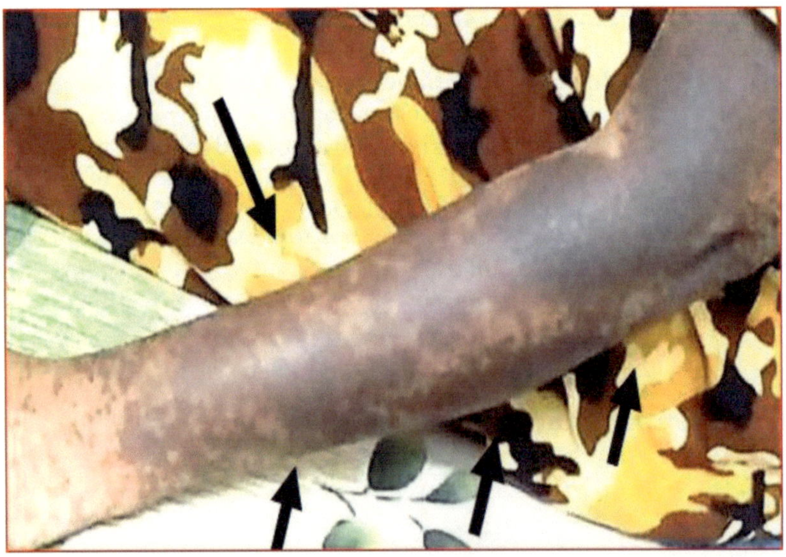

Figure 2.
Epidermal involvement of limbs.

to as necrolysis due to excessive apoptotic keratinocytes [18]. Skin and mucous membranes were involved but with very little inflammation in the dermis referred to as the phenomenon of "dermal silence" by the dermatologist. This was an acute, rare, life-threatening mucocutaneous disease with an annual incidence of approximately 0.4–1.2 cases per million individuals with a mortality rate of more than 30% [19]. SCAR and EuroSCAR pooled data analysis was performed for children under 15 years of age that revealed that anti-bacterials class of drugs such as sulphonamides, anti-epileptics such as phenobarbitol, lamotrigine, and carbamazepine was found to be strongly associated with SJS/TEN in this pediatric population [20]. Following is the case of toxic epidermal necrolysis involving the face and neck (**Figure 1**), trunk, and all four limbs (**Figure 2**) in an elderly female on cefixime and paracetamol treatment for dengue fever with no pre-existing comorbidities.

4. Criteria for the diagnosis of SJS/TEN based on extent of epidermal detachment

Steven Johnson Syndrome is an immune complex-mediated hypersensitivity reaction that involves mucocutaneous body areas such as oral, nasal, ocular, vaginal, urethral, gastrointestinal, and lower respiratory tract mucous membranes. Moreover, gastrointestinal and lower respiratory tract infections may progress to necrosis. The diagnosis of steven johnson syndrome, Toxic epidermal necrolysis, or overlap syndrome totally depends on the afflicted skin body surface area with epidermal detachment (**Figure 3**) [21].

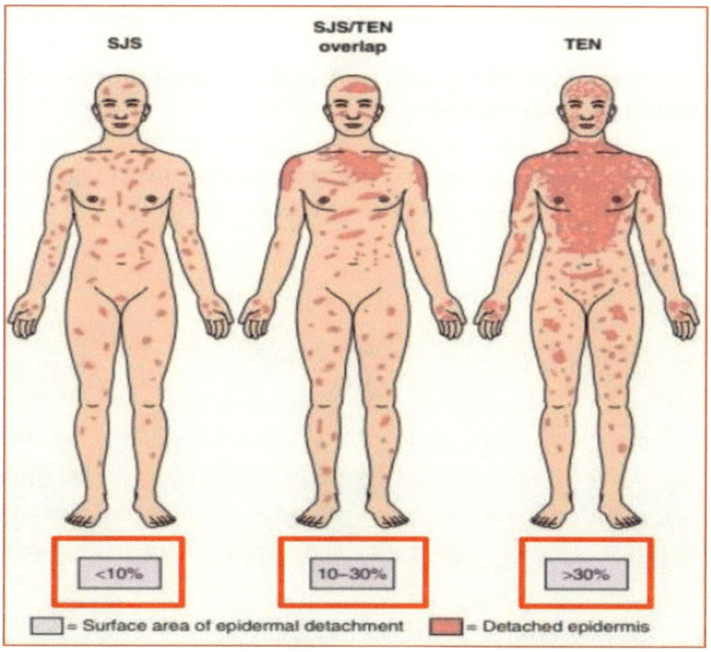

Figure 3.
The extent of epidermal detachment in SJS/TEN. (adapted from fig 21.9 Bolognia and Bastuji-Garin S. et al. arch Derm 129: 92, 1993).

Classification	Bullous Erythema multiforme (EM)	Steven Johnson Syndrome (SJS)	Overlap-Steven Johnson syndrome	Toxic epidermal necrolysis (TEN) with spots	Toxic epidermal necrolysis (TEN) without spots
Epidermal Detachment	<10% BSA	<10% BSA	10–30% BSA	>30% BSA	>10% BSA
Typical Targets	Yes	No	No	No	No
Atypical Targets	Raised	Flat	Flat	Flat	No
Spots	No	Yes	Yes	Yes	No

Abbreviations: BSA- Body Surface Area, EM- Erythema Multiforme, SJS- Steven Johnson Syndrome, TEN - toxic epidermal necrolysis. Adapted from Bastuji-Garin S et al. Clinical classification of cases of toxic epidermal necrolysis, Stevens-Johnson syndrome, and erythema multiforme. Arch Dermatol 1993; 129(1): 92–96 (adapted from Bastuji-Garin S et al.).

Table 2.
Distinguishing features of different severe bullous skin lesions.

- Steven Johnson syndrome (SJS)- < 10% of the body surface area involvement
- SJS/TEN overlap syndrome- 10 – 30% of the body surface area involvement
- Toxic Epidermal Necrolysis (TEN)- >30% of the body surface area involvement

Distinguishing features of different severe bullous skin lesions are tabulated below (**Table 2**) [22].

5. Proposed immunological models of T cell activation in Stevens-Johnson syndrome/toxic epidermal necrolysis

There are three proposed models of T cell activation in Stevens-Johnson syndrome/toxic epidermal necrolysis due to drugs (**Figure 4**) [23].

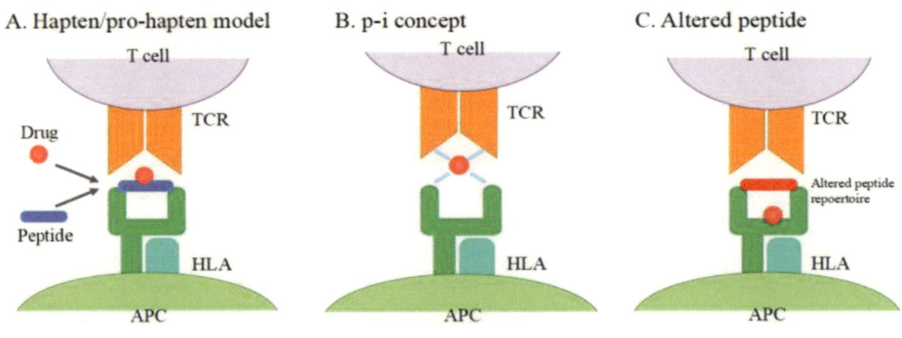

Figure 4.
Models of T cell activation in Steven Johnson syndrome and toxic epidermal necrolysis. Abbreviations: APC- antigen presenting cells, HLA- human leukocyte antigen, TCR- T cell receptor, HLA- human leukocyte antigen. Adapted from: Hasegawa a and Abe R. recent advances in managing and understanding Stevens-Johnson syndrome and toxic epidermal necrolysis (version 1; peer review: 2 approved). F1000Research 2020, 9(F1000 faculty rev):612.

A. Hapten/pro-hapten model: drugs/drug metabolites form complex with carrier proteins and are presented as haptenated peptides in the peptide-binding groove of HLA molecules (covalent).

B. p-i concept: drugs directly bind to HLA & TCR non-covalently.

C. Altered peptide model: drugs bind to the peptide-binding groove of HLA, resulting in alteration of HLA-binding peptide repertoire.

6. Pathophysiology of occurrence of SJS/TEN

1. Drug induced upregulation of Fas-ligand by keratinocytes expressing Fas lead to activation of death receptor-mediated apoptotic pathway [24, 25].

2. Drug interaction with MHC class I-expressing cells leads to drug-specific CD8 + cytotoxic T cells accumulating within the epidermal blisters releasing perforin & granzyme B that kill keratinocytes [26].

3. Drug-activated monocytes secrete annexin A1, which induces necroptosis in keratinocytes [11].

4. Drug triggers activation of CD8 + T cells, NK (Non-Killer) T cells to secrete granulysin leading to keratinocyte death without the need for cell contact [27].

6.1 Histopathologic features

Histopathologic features in SJS/TEN show apoptotic keratinocytes present individually and in clusters within the epidermis. It may also show dendritic cell infiltration, spongiform superficial epidermal pustules, edema of the papillary dermis, and perivascular infiltrates of lymphocytes (**Figure 5**). Subtle vacuolar changes along the basal layer are accompanied by minimal inflammation with scattered lymphocytes within the epidermis [28].

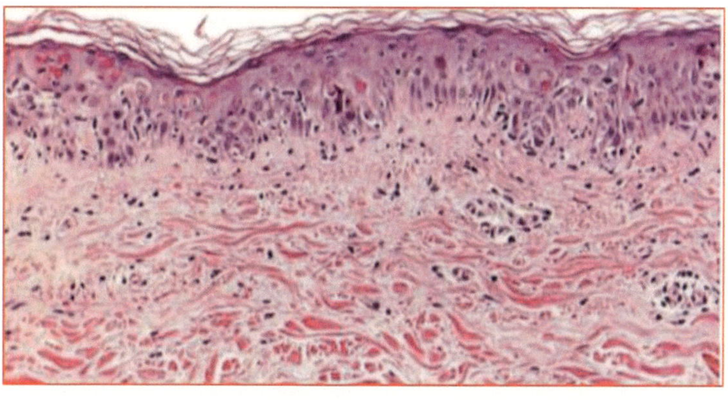

Figure 5.
Histopathologic features in SJS/TEN show apoptotic keratinocytes present individually and in clusters within the epidermis.

6.2 SCORTEN scale (SCORe of toxic epidermal necrosis) for predicting mortality risk in SJS/TEN

SCORTEN scale is the severity-of-illness scale that measures the severity of certain bullous conditions and can be helpful in systematically determining the disease/reaction prognosis. The table below (**Table 3**) provides details of seven independent risk factors that need to be evaluated within the first 24 hours of admission or within the first 5 days of reaction that will add to accuracy in predicting mortality. The presence of prognostic factors is given a score of 1 and if not present then assign the score of 0. SCORTEN score is the summation of these individual scores allotted to each of the seven prognostic factors based on their presence or absence that gives the overall prognosis in terms of percentage mortality. More risk factors present, the higher the SCORTEN score, and the higher the mortality rate [29].

Serial number	Prognostic Factors	Score	SCORTEN score	% Mortality rate
1.	Age > 40 years	1	0–1	3
2.	Tachycardia >120 bpm	1	2	12
3.	Neoplasia	1	3	35
4.	Initial detachment >10%	1	4	58
5.	Serum Urea>10 mmol/l	1	>4	90
6.	Serum Bicarbonate<20 mmol/l	1		
7.	Blood glucose >14 mmol/l	1		

Tablc 3.
Seven independent risk factors for calculating SCORTEN score and predicting mortality risk in SJS/TEN.

7. Algorithm for drug causality for epidermal necrolysis (AlDeN) scale for causality assessment

Table 4 is summarized below.

Criteria	Values	Rules to apply	Value range
Delay from initial drug component intake to onset of reaction (index day)	Suggestive +3	From 5 to 28 days	-3 to 3
	Compatible +2	From 29 to 56 days	
	Likely +1	From 1 to 4 days	
	Unlikely −1	>56 days	
	Excluded −3	Drugs started on or after the index day	
	In case of previous reaction to the same drug only changes for – Suggestive +3: from 1 to 4 days Likely +1: from 5 to 56 days		

Criteria	Values	Rules to apply	Value range
Drug present in the body on index day	Definite 0	Drug continued up to index day or stopped at a time point less than five times the elimination half-life before the index day	-3 to 0
	Doubtful −1	Drug stopped at a time point prior to the index day by more than five times the elimination half-lifea but liver or kidney function alterations or suspected drug interactions are present	
	Excluded −3	Drug stopped at a time point prior to the index day by more than five times the elimination half-life, without liver or kidney function alterations or suspected drug interactions	
Prechallenge/ rechallenge	Positive specific for disease and drug: 4	SJS/TEN after use of same drug	-2 to 4
	Positive specific for disease or drug: 2	SJS/TEN after use of similar drug or other reaction with same drug	
	Positive unspecific: 1	Other reaction after use of similar drug	
	Not done/ unknown: 0	No known previous exposure to this drug	
	Negative −2	Exposure to this drug without any reaction (before or after reaction)	
Dechallenge	Neutral 0	Drug stopped (or unknown)	-2 to 0
	Negative −2	Drug continued without harm	
Type of drug (notoriety)	Strongly associated 3	Drug of the "high-risk" list according to previous case–control studies	-1 to 3
	Associated 2	Drug with definite but lower risk according to previous case–control studies	
	Suspected 1	Several previous reports, ambiguous epidemiology results (drug "under surveillance")	
	Unknown 0	All other drugs including newly released ones	
	Not suspected −1	No evidence of association from previous epidemiology study with sufficient number of exposed controls	
	Intermediate score = total of all previous criteria		−11 to 10
Other cause	Possible −1	Rank all drugs from highest to lowest intermediate score	−1

Criteria	Values	Rules to apply	Value range
		If at least one has an intermediate score > 3, subtract 1 point from the score of each of the other drugs taken by the patient (another cause is more likely)	
Final score − 12 to 10			
Causality assessment scale <0, Very unlikely; 0–1, unlikely; 2–3, possible; 4–5, probable; ≥6, very probable.			

Table 4.
Details of AlDeN scale for causality assessment (algorithm for drug causality for epidermal necrolysis).

8. Drug Reaction with Eosinophilia and Systemic Symptoms (DRESS)/ Drug-induced Hypersensitivity Syndrome (DIHS)

This DRESS is a life-threatening drug hypersensitivity reaction that can manifest with fever, cutaneous eruptions, and internal organ involvement [30]. Though the disease has a 10% mortality risk, it involves more than 50% of body surface area with skin lesions showing infiltrative papules and plaques with purpuric changes. This may be associated with other clinical features such as facial edema, desquamation during the resolution stage, and mucosal lesion involving the mouth and lips most commonly. Systemic symptoms depend on the organ involved. Eosinophilia (66–95%) is the most common hematological abnormality noted followed by atypical lymphocytosis (27–67%), lymphadenopathy (54%), and the liver is the most common internal organ involved followed by kidney, lung, and heart. The onset of skin reactions is usually observed after 3–8 weeks of the start of suspect drugs. The drugs implicated in such drug reactions are anti-convulsants, anti-infectious agents (anti-tuberculosis, antibiotics, and antiviral agents), sulfonamides, and uric acid-lowering medications. Usually, DRESS lasts for more than 15 days with prolonged courses with flare-ups observed after Human Herpes Virus type 6 reactivation. The immunopathology blamed here is the antiviral and anti-drug immune responses contribute to the disease presentations but the exact interaction is still unclear. Sequelae of this disease could be fulminant type 1 diabetes mellitus, thyroid disorders, autoimmune diseases, or permanent renal dysfunction that needs dialysis [31].

9. Acute generalized exanthematous pustulosis (AGEP)

The clinical characteristics of Acute Generalized Exanthematous Pustulosis (AGEP) are typical with multiple sterile, non-follicular pustules on oedematous erythema mostly on major skin folds. It may be associated with facial edema, blisters, or atypical target lesions. Mucosal lesions are rare and mild. There may be fever and leukocytosis associated with cutaneous eruptions. Systemic involvement is rare (<20%) but may affect the liver followed by the kidney, lung, and bone marrow. The culprit drugs that are strongly associated with these reactions are pristinamycin, ampicillin/amoxicillin, quinilones, hydroxychloroquine, anti-infective sulfonamides, terbinafine, and diltiazem based on multinational case–control EuroSCAR study. The duration between drug intake and occurrence of skin eruption may take a median duration of 1 day in the case of antibiotics and 11 days for other medications. AGEP resolves without any sequelae and has a good prognosis as well [32]. The diagnostic criteria for Acute Generalized Exanthematous Pustulosis (AGEP) are as follows [33, 34] – 1) Acute pustular eruption 2) Fever >38 degrees Celsius 3) Neutrophilia with or without eosinophilia 4) Sub corneal or intradermal pustules on biopsy 5) Spontaneous resolution in less than 15 days.

10. Diagnostic tests for cutaneous drug hypersensitivity

Drug associated cutaneous hypersensitivity reaction requires an adequate clinical and physical examination approach with an evaluation of historical details. In addition to clinical history and examination, the diagnostic test will help to confirm the diagnosis of SCARs. Also, it will help us to determine if the initial reaction is IgE or non-IgE mediated. A diagnostic test can be classified into In Vitro tests and In Vivo tests.

In Vitro Tests: The quantification of various chemokines (histamine, tryptase, leukotrienes) from a different sample such as peripheral blood, nasal or bronchial secretions, urine sample serves as useful means to differentiate between the immediate and delayed types of hypersensitivity reactions (For details, refer **Table 5**). The Allergen challenge test may be formed to see the difference in the level of chemical mediators from the baseline levels. In case of acute anaphylactic reactions, the test can serially measure serum total tryptase levels, at 1 and 6 hours but is not completely reliable as normal levels may be detected in fatal cases of anaphylaxis. Even plasma histamine levels drop after 1 hour of anaphylactic symptoms making this test not reliable. Moreover, these test kits are expensive.

In Vivo Tests: Many skin tests are available as useful tools in the diagnosis of IgE-mediated allergy. A skin prick test (SPT) is positive if the mean wheal diameter is≥3 mm after 15 to 20 minutes (associated with a flare response) compared to the negative control. Similarly, the intradermal test (IDT) is positive when meaning wheal diameter is ≥3 mm compared to the baseline diameter for the negative control after 15 to 20 minutes of the intradermal administration of allergen (0.02 to 0.05 ml). Though IDT is more sensitive than the skin prick test, there is still a higher risk of irritation, false positive reaction, and IgE-dependent anaphylactic reactions. The observation for immediate hypersensitivity reaction usually appears in 15 to 20 minutes while 24 to 72 hours' evaluation is required for non-immediate (late) reactions. Patch tests are used for the diagnosis of delayed hypersensitivity drug reactions. In these tests, a patch embedded with the suspected allergen is fixed on the back of the patient for 1 to 2 days and the result is read after 1 day and/or after 2 to 3 days. A photo-patch test is a modified patch test used to diagnose suspected photoallergic or phototoxic reactions. This patch is removed after a day and a skin area of almost 10 J/cm2 is irradiated with ultraviolet A light and the results are read on days 2, 3, and 4. The use of non-irritating skin test concentrations is recommended for SPT, IDT, and patch tests. In very rare cases, a skin biopsy may be helpful in differential diagnoses of skin conditions with typical histological patterns. For instance, connective tissue disease is characterized by interface dermatitis with epidermal atrophy, focal parakeratosis, dermal mucinosis, and fibrinoid deposition in the dermis [35, 36]. Drug provocation (challenge) tests (DPTs) are performed using suspected agents and objectively reproduce the patient's symptoms and signs of hypersensitivity. However, the positive test does not confirm an immune-mediated reaction. For DPTs, the drug is given using slow, incremental dose escalations at fixed time intervals and observing for the presence or absence of an objective reaction under the strict supervision of trained clinicians/nurses with

Function of type of in vitro test	Test name	Drug names/Disease Conditions	Limitation of the test/ Inferences
Allergen-specific IgE levels	radioallergosorbent tests (RASTs) or radioimmunoassay (RIA) or fluorescent enzyme immunoassay (FEIA) tests (ImmunoCAP®)	penicilloyl, amoxicilloyl, ampicilloyl, cefaclor, protamine, insulin, suxamethonium, neuromuscular blocking agents (NMBAs), and chlorhexidine	specific, lack sensitivity compared to clinical history and/or skin tests. (in clinical practice, a utility for these tests is limited).
CD 63 and CD 203c levels on activated basophils.	Flow cytometry-based basophil activation assays (also known as flow cellular antigen stimulation tests [CASTs])	beta-lactam, NSAIDs, fluoroquinolones, iodinated contrast media, proton pump inhibitors, NMBAs, and chemotherapeutical agents	technical concerns, false-positive results, and lack of sensitivity and specificity. (not widely used)
Peripheral blood eosinophilia or elevated total IgE levels- presence or absence.	Complete Hemogram profile and Serum IgE levels in the blood	drug hypersensitivity syndromes.	not useful in the diagnosis or exclusion of drug allergies.
Positive direct Coomb's test		hematological manifestations of drug hypersensitivity e.g. immune-mediated haemolytic anemia, leukopaenia, thrombocytopaenia	no specific diagnostic test or serological test apart from recovery of the cytopaenia following the withdrawal of the putative drug, Drug-induced IgM and IgG have not been found to be clinically useful
Lymphocyte transformation test (LTT)	diagnosis of reactions in a wide variety of delayed reactions with a wide variety of drugs	T-cell mediated delayed hypersensitivity against a wide variety of drugs	positive LTT is useful in confirming the diagnosis, a negative test cannot exclude drug hypersensitivity. Positive LTT is usually drug-specific, and reaction-specific.

Table 5.
The following table provides information about the in-vitro tests.

well-equipped resuscitative tools. The DPT can help to exclude drug hypersensitivity when the history is nonsuggestive or the symptoms nonspecific. To definitively diagnose drug allergy when the clinical history is suggestive but allergological tests are negative, inconclusive, or unavailable. Specific contraindications to DPT include pregnancy; comorbidities in which DPT may provoke the medical situation beyond the ability to control it (e.g., acute infections; uncontrolled asthma; or underlying cardiac, hepatic, or renal diseases); immunobullous drug eruptions; and cases in which the initial reaction was a severe cutaneous and/or systemic reaction (e.g., SJS and TEN). The risks and benefits of any DPT must be explained to the patient and informed consent obtained. Short-acting antihistamines (e.g., chlorpheniramine or hydroxyzine) should be stopped for 3 days and long-acting antihistamines (e.g., cetirizine, loratadine, or fexofenadine) for 7 days before performing any DPT. Patients should also be fasted overnight and carefully observed at all times during the DPT for symptoms or signs of an adverse reaction. Resuscitation equipment should be available at all times, and staff should be trained in the management of acute anaphylaxis [37].

11. Management of Severe Cutaneous Adverse Reactions

11.1 Treatment of SJS/TEN

11.1.1 General management/supportive care

Correct identification of the causative drug and immediate withdrawal of potentially causative drugs. May use the ALDEN algorithm to determine the culprit drug. Supportive management of skin wounds with anti-shear dressings, nutrition status, electrolyte balance, renal and airway function, and adequate pain control, prevent or treat the wound infections. Refer to the specialized unit/burn center for supportive care, silver wraps, apply emollients, air fluidized beds, and prevent infections. Fluid balance is very important to prevent end-organ hypo perfusion with daily monitoring of urine output (maintain 0.5–1.0 ml/kg/hr) or intra-arterial hemodynamic monitoring. Adequate nutrition supplement for protein loss- 20-25 kcal/kg/day in the early phase and 25–30 kcal/kg/day in the recovery phase is recommended either through oral intake or nasogastric feed. Analgesic care with acetaminophen in mild cases or opiod-based analgesics based on the severity of pain [38, 39].

11.1.2 Specific treatment for SJS/TEN

Corticosteroids- Systemic corticosteroids have shown non-inferiority when compared with supportive care in treating patients with SJS. However, in cases of TEN, many studies have shown survival benefits to the patient but some studies have also shown a lack of efficacy and also an increase in mortality. High doses of systemic steroids have shown to be more effective in TEN patients as recommended by Japanese experts. Araki *et al* have successfully used corticosteroid pulse therapy (methylprednisolone 500 mg/day for 3 days in 5 patients of TEN and all survived also supported by one recent published meta-analysis suggesting systemic corticosteroids as promising immunomodulating therapies for SJS/TEN [40].

Intravenous Immunoglobulin- some studies have shown the survival benefit of Intravenous immunoglobulins (IVIG) with a dose of 2.8 g/kg up to 4 g/kg [41, 42]. Those studies with no survival benefit used doses mostly up to 2 g/kg or lower [43]. Huang et al. performed the first meta-analysis on the efficacy of IVIg for the treatment of TEN showing the benefit of the use of high dose IVIG over low dose IVIG [44]. But recent published reviews and meta-analyses showed no differences in mortality with IVIG compared with only supportive care [45, 46].

Cyclosporine- Inhibits CD8+ cytotoxic T cells with anti-apoptotic effect by inhibition of Fas ligand. Cyclosporine 3 mg/kg for 10 days with gradual tapering over 1 month revealed less skin detachment and a lower mortality rate in one of the pilot studies recruiting 29 patients of SJS/TEN. Chen *et al* in a meta-analysis found significantly lower mortality than calculated as per SCORTEN score in patients receiving cyclosporine (OR: 0.42, 95% CI- 0.19-0.95) [47, 48]. Zimmermann et al. meta-analysis also showed a better reduction in mortality with the use of cyclosporine when compared with just supportive care. However large-scale randomized controlled studies are required to confirm these findings [49].

Anti-TNF alpha agents: Increase in TNF-alpha in skin specimens, skin blister fluids, and in serum of SJS/TEN patients have been observed and hence the role of anti-TNF alpha agents may prove effective. Infliximab and Etanercept use have shown to be effective as published in many case reports and case series with better survival outcomes. Only one trial by Wolkenstein *et al* that used Thalidomide for treatment of SJS/TEN was prematurely terminated in view of the increase in mortality observed in patients [50, 51].

Plasmapheresis: Plasmapheresis is known to filter the harmful mediators in blood and have shown dramatic improvement in a patient with SJS/TEN. Narira *et al* study demonstrated that the use of plasmapheresis in patients refractory to conventional therapy treatment reduced the interleukin levels (IL-6, 8, and TNF-alpha) and is recommended for use by Japanese experts in TEN patients refractory to high dose corticosteroids [52].

11.2 Treatment for drug reaction with eosinophilia and systemic symptoms (DRESS) or drug-induced hypersensitivity syndrome (DIHS)

- Supportive care is required with hydration maintenance in most of the cases except those with detected elevated levels of auto-antibodies.

- Systemic corticosteroids such as prednisolone with starting dose of 0.5 to 1 mg/kg/day gradually tapering over 2–3 months suggested by some experts known to decrease flare-up episodes and auto-immune sequelae.

- However, there is a risk for higher rates of infections, and septicemia and may require intensive care management. Hence suggested use only in those with severe presentations.

- French group of Dermatology recommends the use of systemic use of corticosteroids for those with a 5-fold increase in serum transaminases level or if there is the involvement of another organ such as kidney, lung, and heart.

- The results of the use of IVIG are conflicting and its use as monotherapy should be avoided. Several immunosuppressants like cyclosporine, cyclophosphamide, mycophenolate mofetil, and rituximab have been proposed in addition to systemic corticosteroids or IVIG in patients with severe disease and viral reactivation (Human Herpes virus –6 activation has been implicated as a trigger for this reaction) [33].

11.3 Acute generalized Exanthematous Pustulosis (AGEP)

- Identification and removal of the culprit drugs are sufficient and skin lesions resolve in 6–8 days after suspect drug withdrawal [53].

- Hospitalization may be required in some patients and may be treated with topical and systemic corticosteroids if required. The beneficial effects of systemic steroids need further evaluation.

11.4 Generalized bullous fixed drug eruptions (GBFDE)

- Prompt identification and withdrawal of causative suspect drugs remains the mainstay management of this reaction [54].

- Systemic corticosteroids may be beneficial in severe cases though there is a lack of sufficient evidence to compare with other treatment modalities and supportive care.

12. Future directions using pharmacogenetic tools to identify genetic predisposition to SCARs

The occurrence of SCAR is unpredictable but this uncertainty can be minimized to some extent by performing a pharmacogenetic assessment. If done before initiating patients on potential drugs, the occurrence of SCAR can be prevented as well. Human Leukocyte Antigens (HLA) genes have been evaluated in detail that has been found to have an association with SCAR variations observed in some patients on specific drugs (listed in **Table 6**). Having knowledge about genetic predisposition will help to detect patients at higher risk of developing SCAR. A study by Esmaeilzadeh H. *et al* study from Iran has shown Steven Johnson Syndrome (SJS)/Toxic Epidermal Necrolysis (TEN) as the most common drug-induced SCAR presentation and the most common culprit drug as beta-lactam antibiotics followed by carbamazepine. The presence of HLA-A*02:01 and A*51:01 was also shown to increase the risk of SCAR while A*11:01 had a protective role against SCAR. Those with HLA-A*02:01, HLA-A*24:02, and HLA-B*51:01 were an increased SJS as observed in the same study [55].

Predisposing Infectious diseases	Drug-induced	Malignancy related	HLA allele associated with the risk to specific drugs
AIDS (Acquired Immunodeficiency syndrome)	Antiepileptics: Phenytoin, Carbamazepine, Oxcarbazepine, Valproic acid, Lamotrigine	Hematological	HLA-B*15:02- Carbamazepine
Coxsackie and Dengue virus infection	Barbiturates	Lung cancer	HLA-A*31:01- Carbamazepine
Influenza, Herpes Simplex virus	Anti-retrovirals: NRTI- Nevirapine NNRTIs - Indinavir	Malignant Lymphoma	HLA-B*57:01- Allopurinol Abacavir
Bacterial Infection: Gr A beta-hemolytic streptococci, Diphtheria, Brucellosis, LGV, Mycobacteria	NSAIDS: oxicams –meloxicam, piroxicam, Paracetamol	Urothelial Carcinoma	HLA-A*32:01 Vancomycin
Fungal infection: coccidiomycosis, dermatophytosis, trichomoniasis	TNF alpha antagonist: Etanercept, Infliximab, Adalimumab	Hepatocellular Carcinoma	

Abbreviations: AIDS- Acquired immune deficiency syndrome, LGV- Lymphogranuloma venereum, NRTI- Nucleoside reverse transcriptase inhibitors, NNRTIs- Non-nucleoside reverse transcriptase inhibitors, NSAIDs- Non-steroidal anti-inflammatory drugs, TNF alpha- Tumor Necrosis Factor-alpha, HLA- Human leukocyte antigens.

Table 6.
Predisposing factors for Steven Johnson syndrome/toxic epidermal necrolysis.

13. Conclusion

Among the various pathophysiology suggested for SCARs, T cells play a major role in the occurrence of delayed hypersensitivity reactions presenting as SCAR variants. There is a need to select structurally different classes of antibiotics in case of antibiotic-induced severe cutaneous adverse reaction (SCAR) to avoid recurrence. Furthermore, there is a need to conduct research and explore the immune mechanism of viral-drug-gene interaction and develop drugs to modulate T cells/ other cell lineages/novel therapeutic targets. Animal models for SCAR variants may be an important step toward the development of new drugs but are understudied. Nevertheless, rational use of antibiotics should be promoted as well as implemented into practice by consumers and health care physicians. The important concern with Immunomodulators and targeted therapies is associated with long-term sequelae such as immune reactive inflammatory syndrome and polyglandular autoimmune syndrome III. Validation of more prognostic HLA and other biomarkers for guiding therapy may help to prevent the occurrence of scars. There is a need to develop multicentric, multi-ethnic, and multi-regional registries that will enable us to gather clinical and demographic profiles, along with histopathological, genomic, proteomic, and metabolomic evaluation supplemented with a data mining approach for a better understanding of signals generated. The metabolomic approach may be helpful to detect the drug or its metabolite may be the causative factor for SCAR severity.

Abbreviations

SCARs	Severe Cutaneous Adverse Drug Reactions
AGEP	Acute Generalized Eaxnthematous Pustulosis
SJS	Steven Johnson Syndrome
TEN	Toxic Epidermal Necrolysis
SJS/TEN Overlap	Steven Johnson Syndrome/Toxic Epidermal Necrolysis Overlap syndrome
GBFDE	Generalized Bullous Fluid Drug Eruptions
EEMM	Erythema Exsudativum Multiforme with Mucosal Involvement
EM	Erythema Multiforme
DRESS	Drug Related Eosinophilia with Eosinophilic Systemic Symptom
WAO	World Allergy Organization
DIHS	Drug Induced Hypersensitivity Syndrome
BC	Before Christ
CD	Clusters of Differentiation
Ig E	Immunoglobulin E
Ig G	Immunoglobulin G
Ig M	Immunoglobulin M
HP	Hypersensitivity Pneumonitis
PAN	Polyarthritis Nodosa
SLE	Systemic Lupus Erythematosus
PPD	Purified Protein Derivative
ADRs	Adverse Drug Reactions

NSAIDs	Non-Steroidal Anti-Inflammatory Drugs
RIA	Radioimmunoassay
TCR	T Cell Receptor
APC	Antigen Presenting Cells
FEIA	Fluorimetric Enzyme-Linked Immunoassay
SPT	Skin Prick Test
OR	Odds Ratio
IDT	Intradermal test
IVIG	Intravenous Immunoglobulin
TNF	alpha- Tumor Necrosis Factor-alpha
HLA	Human Leukocyte Antigen
AIDS	Acquired Immunodeficiency Syndrome
LGV	Lymphogranuloma Venerum
NRTI	Nucleoside Reverse Transcriptase Inhibitors
NNRTI	Non-Nucleoside Reverse Transcriptase Inhibitors
LTT	Lymphocyte Transformation Test
DPT	Drug Provocation Tests
CASTs	Cellular Antigen Stimulation Tests

Author details

Miteshkumar Rajaram Maurya[1*], Renuka Munshi[1], Sachin Bhausaheb Zambare[2] and Sanket Thakur[3]

1 Department of Clinical Pharmacology, Topiwala National Medical College and BYL Nair Charitable Hospital, Mumbai, Maharashtra, India

2 DNB Nephrology, Sterling Hospital, Vadodara, Gujarat, India

3 Pediatric Critical Care, Royal Manchester Children's Hospital, Manchester, United Kingdom

*Address all correspondence to: mitesh.maurya4@gmail.com

References

[1] Adler NR, Aung AK, Ergen EN, Trubiano J, Goh MSY, Phillips EJ. Recent advances in the understanding of severe cutaneous adverse reactions. The British Journal of Dermatology. 2017;**177**(5):1234-1247

[2] Bellón T. Mechanisms of severe cutaneous adverse reactions: Recent advances. Drug Safety. 2019;**42**(8): 973-992

[3] Shafiq N, Bhattacharjee S, Malhotra S. Severe cutaneous adverse reaction (SCAR). Indian Dermatology Online Journal. 2022;**13**(1):10-12

[4] Patel RM, Marfatia YS. Clinical study of cutaneous drug eruptions in 200 patients. Indian Journal of Dermatology, Venereology and Leprology. 2008;**74**:80

[5] Classification of SCAR. Available from: https://www.worldallergy.org/education-and-programs/education/allergic-disease-resource-center/professionals/drug-allergies. [Accessed: June 22, 2022]

[6] Fierer J, Looney D, Pechère JC. Nature and pathogenicity of micro-organisms. Infectious Diseases. 2017:4-25. e1. Available from: https://www.ncbi.nlm.nih.gov/pmc/articles/PMC7173513/

[7] Wiffen P, Gill M, Edwards J, Moore A. Adverse drug reactions in hospital patients. A systematic review of the prospective and retrospective studies. Bandolier Extra. 2002;**2**:1-16

[8] Gruchalla R. Understanding drug allergies. The Journal of Allergy and Clinical Immunology. 2000;**105**:637-644

[9] Schwartz RA, McDonough PH, Lee BW. Toxic epidermal necrolysis. Journal of the American Academy of Dermatology. 2013;**69**(2):187.e1-187.e16

[10] Roujeau JC, Bastuji-Garin S. Systematic review of treatments for Stevens-Johnson syndrome and toxic epidermal necrolysis using the SCORTEN score as a tool for evaluating mortality. Ther Adv Drug Saf. 2011;**2**:87-94

[11] Roujeau JC. Clinical heterogeneity of drug hypersensitivity. Toxicology. 2005;**209**(2):123-129

[12] Diphoorn J, Cazzaniga S, Gamba C, Schroeder J, Citterio A, Rivolta AL, et al. Incidence, causative factors and mortality rates of Stevens-Johnson syndrome (SJS) and toxic epidermal necrolysis (TEN) in northern Italy: Data from the REACT registry. Pharmacoepidemiology and Drug Safety. 2016;**25**:196-203

[13] Hsu DY, Brieva J, Silverberg NB, Silverberg JI. Morbidity and mortality of Stevens-Johnson syndrome and toxic epidermal necrolysis in United States adults. The Journal of Investigative Dermatology. 2016;**136**:1387-1397

[14] Kang DY, Yun J, Lee SY, Koh YI, Sim DW, Kim S, et al. A Nationwide study of severe cutaneous adverse reactions based on the multicenter registry in Korea. The Journal of Allergy and Clinical Immunology. In Practice. 2021;**9**(2):929-936.e7

[15] Current status of Phase I of the RegiSCAR project. 2014. Available from: http://www.regiscar.org/Status.html. [Accessed: June 22, 2022]

[16] Stevens AM, Johnson FC. A new eruptive fever associated with stomatitis and ophthalmia: Report of two cases in

children. American Journal of Diseases of Children. 1922;**24**:526-533

[17] Callahan SW, Oza VS. Stevens-Johnson syndrome—A look Back. JAMA Dermatology. 2017;**153**(2):240

[18] Schwartz RA, McDonough PH, Lee BW. Toxic epidermal necrolysis: Part I. introduction, history, classification, clinical features, systemic manifestations, etiology, and immunopathogenesis. Journal of the American Academy of Dermatology. 2013;**69**(2):173.e1-173.13 quiz 185-6

[19] Levi N, Bastuji-Garin S, Mockenhaupt M, et al. Medications as risk factors of Stevens-Johnson syndrome and toxic epidermal necrolysis in children: A pooled analysis. Pediatrics. 2009;**123**(2):e297-e304

[20] Bastuji-Garin S, Rzany B, Stern RS, Shear NH, Naldi L, Roujeau JC. Clinical classification of cases of toxic epidermal necrolysis, Stevens-Johnson syndrome, and erythema multiforme. Archives of Dermatology. 1993;**129**(1):92-96

[21] Harr T, French LE. Toxic epidermal necrolysis and Stevens-Johnson syndrome. Orphanet Journal of Rare Diseases. 2010;**5**:39

[22] Hasegawa A, Abe R. Recent advances in managing and understanding Stevens-Johnson syndrome and toxic epidermal necrolysis. F1000Res. 2020;**9**:612

[23] Itoh N, Yonehara S, Ishii A, Yonehara M, Mizushima S, Sameshima M, et al. The polypeptide encoded by the cDNA for human cell surface antigen Fas can mediate apoptosis. Cell. 1991;**66**:233-243

[24] Iwai K, Miyawaki T, Takizawa T, Konno A, Ohta K, Yachie A, et al. Differential expression of bcl−2 and susceptibility to anti-Fas-mediated cell death in peripheral blood lymphocytes, monocytes, and neutrophils. Blood. 1994;**84**:1201-1208

[25] Inachi S, Mizutani H, Shimizu M. Epidermal apoptotic cell death in erythema multiforme and Stevens-Johnson syndrome. Contribution of perforin-positive cell infiltration. Archives of Dermatology. 1997;**133**:845-849

[26] Saito N, Qiao H, Yanagi T, et al. An annexin A1-FPR1 interaction contributes to necroptosis of keratinocytes in severe cutaneous adverse drug reactions. Science Translational Medicine. 2014;**6**(245):245-295. DOI: 10.1126/scitranslmed.3008227

[27] Rzany B, Hering O, Mockenhaupt M, Schröder W, Goerttler E, Ring J, et al. Histopathological and epidemiological characteristics of patients with erythema exudativum multiforme major, Stevens-Johnson syndrome and toxic epidermal necrolysis. The British Journal of Dermatology. 1996;**135**(1):6-11

[28] Bastuji-Garin S, Fouchard N, Bertocchi M, Roujeau JC, Revuz J, Wolkenstein P. SCORTEN: A severity-of-illness score for toxic epidermal necrolysis. The Journal of Investigative Dermatology. 2000;**115**(2):149-153. DOI: 10.1046/j.1523-1747.2000.00061.x

[29] Sassolas B, Haddad C, Mockenhaupt M, Dunant A, Liss Y, Bork K, et al. ALDEN, an algorithm for assessment of drug causality in Stevens-Johnson syndrome and toxic epidermal necrolysis: Comparison with case-control analysis. Clinical Pharmacology and Therapeutics. 2010;**88**(1):60-68. DOI: 10.1038/clpt.2009.252

[30] Choudhary S, McLeod M, Torchia D, Romanelli P. Drug reaction with eosinophilia and systemic symptoms (DRESS) syndrome. The

Journal of Clinical and Aesthetic Dermatology. 2013;**6**(6):31-37

[31] Feldmeyer L, Heidemeyer K, Yawalkar N. Acute generalized Exanthematous Pustulosis: Pathogenesis, genetic background, clinical variants and therapy. International Journal of Molecular Sciences. 2016;**17**(8):1214. DOI: 10.3390/ijms17081214

[32] Sidoroff A, Halevy S, Bavinck JN, Vaillant L, Roujeau JC. Acute generalized exanthematous pustulosis (AGEP)--a clinical reaction pattern. Journal of Cutaneous Pathology. 2001;**28**:113-119

[33] De A, Das S, Sarda A, Pal D, Biswas P. Acute generalised Exanthematous Pustulosis: An update. Indian Journal of Dermatology. 2018;**63**(1):22-29. DOI: 10.4103/ijd.IJD_581_17

[34] Brockow K, Przybilla B, Aberer W, et al. Guideline for the diagnosis of drug hypersensitivity reactions: S2K-guideline of the German Society for Allergology and Clinical Immunology (DGAKI) and the German dermatological society (DDG) in collaboration with the Association of German Allergologists (AeDA), the German Society for Pediatric Allergology and Environmental Medicine (GPA), the German contact dermatitis research group (DKG), the Swiss Society for Allergy and Immunology (SGAI), the Austrian Society for Allergology and Immunology (ÖGAI), the German academy of Allergology and environmental medicine (DAAU), the German Center for Documentation of severe skin reactions and the German Federal Institute for Drugs and Medical Products (BfArM). Allergo J Int. 2015;**24**(3):94-105. DOI: 10.1007/s40629-015-0052-6

[35] Bergmann MM, Caubet JC. Role of in vivo and in vitro tests in the diagnosis of severe cutaneous adverse reactions (SCAR) to drug. Current Pharmaceutical Design. 2019;**25**(36):3872-3880. DOI: 10.2174/138161282566619 1107104126

[36] Defrance C, Bousquet PJ, Demoly P. Evaluating the negative predictive value of provocation tests with nonsteroidal anti-inflammatory drugs. Allergy. 2011;**66**:1410-1414. DOI: 10.1111/j.1398-9995.2011.02671.x

[37] Ushigome Y, Kano Y, Ishida T, Hirahara K, Shiohara T. Short- and long-term outcomes of 34 patients with drug-induced hypersensitivity syndrome in a single institution. Journal of the American Academy of Dermatology. 2013;**68**(5):721-728

[38] Cho YT, Chu CY. Treatments for severe cutaneous adverse reactions. Journal of Immunology Research. 2017;**2017**:1503709. DOI: 10.1155/2017/1503709

[39] Auyeung J, Lee M. Successful treatment of Stevens-Johnson syndrome with cyclosporine and corticosteroid. The Canadian Journal of Hospital Pharmacy. 2018;**71**(4):272-275

[40] Scheuerman O, Nofech-Moses Y, Rachmel A, Ashkenazi S. Successful treatment of antiepileptic drug hypersensitivity syndrome with intravenous immune globulin. Pediatrics. 2001;**107**(1):e14

[41] Fields KS, Petersen MJ, Chiao E, Tristani-Firouzi P. Case reports: Treatment of nevirapine-associated dress syndrome with intravenous immune globulin (IVIG) journal of drugs. Dermatology. 2005;**4**(4):510-513

[42] Joly P, Janela B, Tetart F, et al. Poor benefit/risk balance of intravenous immunoglobulins in DRESS. Archives of Dermatology. 2012;**148**(4):543-544.

DOI: 10.1001/archderm.148.4.dlt120002-c

[43] Huang YC, Li YC, Chen TJ. The efficacy of intravenous immunoglobulin for the treatment of toxic epidermal necrolysis: A systematic review and meta-analysis. The British Journal of Dermatology. 2012;**167**(2):424-432

[44] Bachot N, Revuz J, Roujeau J. Intravenous immunoglobulin treatment for Stevens-Johnson syndrome and toxic epidermal necrolysis: A prospective noncomparative study showing No benefit on mortality or progression. Archives of Dermatology. 2003;**139**(1):33-36

[45] Ye LP, Zhang C, Zhu QX. The effect of intravenous immunoglobulin combined with corticosteroid on the progression of Stevens-Johnson syndrome and toxic epidermal necrolysis: A meta-analysis. PLoS One. 2016;**11**(11):e0167120

[46] Chen Y-T, Hsu C-Y, Chien Y-N, Lee W-R, Huang Y-C. Efficacy of cyclosporine for the treatment of Stevens-Johnson syndrome and toxic epidermal necrolysis: Systemic review and meta-analysis. Dermatologica Sinica. 2017;**35**(3):131-137

[47] Ng QX, De Deyn MLZQ, Venkatanarayanan N, Ho CYX, Yeo WS. A meta-analysis of cyclosporine treatment for Stevens-Johnson syndrome/toxic epidermal necrolysis. Journal of Inflammation Research. 2018;**11**:135-142

[48] Zimmermann S, Sekula P, Venhoff M, Motschall E, Knaus J, Schumacher M, et al. Systemic Immunomodulating therapies for Stevens-Johnson syndrome and toxic epidermal necrolysis: A systematic review and meta-analysis. JAMA Dermatology. 2017;**153**(6):514-522. DOI: 10.1001/jamadermatol.2016.5668

[49] Wolkenstein P, Latarjet J, Roujeau JC, Duguet C, Boudeau S, Vaillant L, et al. Randomised comparison of thalidomide versus placebo in toxic epidermal necrolysis. Lancet. 1998;**352**(9140):1586-1589

[50] Zhang S, Tang S, Li S, Pan Y, Ding Y. Biologic TNF-alpha inhibitors in the treatment of Stevens-Johnson syndrome and toxic epidermal necrolysis: A systemic review. The Journal of Dermatological Treatment. 2020;**31**(1):66-73

[51] Yamada H, Takamori K. Status of plasmapheresis for the treatment of toxic epidermal necrolysis in Japan. Therapeutic Apheresis and Dialysis. 2008;**12**(5):355-359

[52] Husain Z, Reddy BY, Schwartz RA. DRESS syndrome: Part II. Management and therapeutics. Journal of the American Academy of Dermatology. 2013;**68**(5):709.e1-709.e9; quiz 718-20. DOI: 10.1016/j.jaad.2013.01.032

[53] Byrd RC, Mournighan KJ, Baca-Atlas M, Helton MR, Sun NZ, Siegel MB. Generalized bullous fixed-drug eruption secondary to the influenza vaccine. JAAD Case Rep. 2018;**4**(9):953-955

[54] Esmaeilzadeh H, Farjadian S, Alyasin S, Nemati H, Nabavizadeh H, Esmaeilzadeh E. Epidemiology of severe cutaneous adverse drug reaction and its HLA association among pediatrics. Iran J Pharm Res. 2019;**18**(1):506-522

[55] Frey N, Jossi J, Bodmer M, Bircher A, Jick SS, Meier CR, et al. The epidemiology of Stevens-Johnson syndrome and toxic epidermal necrolysis in the UK. The Journal of Investigative Dermatology. 2017;**137**:1240-1247